THE
CONTROL
OF
BOILERS

SAM G. DUKELOW

WITHDRAWN

ISA PRESS

THE CONTROL OF BOILERS

Printed in the United States of America.

We would like to thank the many suppliers who provided material for this book, and we regret any we may have inadvertently failed to credit for an illustration. On notification we shall insert a correction in any subsequent printings.

Some material herein has previously appeared in *Improving Boiler Efficiency* by Sam G. Dukelow, produced by Kansas State University and distributed by ISA.

In preparing this work the author and publishers have not investigated or considered patents which may apply to the subject matter hereof. It is the responsibility of the readers and users of the subject matter to protect themselves against liability for infringement of patents. The information and recommendations contained herein are not intended for specific applications but are of a general educational nature. Accordingly, the authors and publishers assume no responsibility and disclaim all liability of any kind, however arising, as a result of using the subject matter of this work.

The equipment referenced in this work has been selected by the author as examples of the technology, no endorsement of any product is intended by the publisher. In all instances, the manufacturer's procedures should prevail regarding the use of specific equipment. No representation, expressed or implied, is made with regard to the availability of any equipment, process, formula or other procedures contained herein.

INSTRUMENT SOCIETY of AMERICA
67 Alexander Drive, P.O. Box 12277
Research Triangle Park, NC 27709

Library of Congress Cataloging-in-Publication Data
Dukelow, Sam G., 1917–
   The control of boilers.

   Includes index.
   1. Steam-boilers — Automatic control.   I. Title.
TJ288.D78   1986   621.1'83   86-20190
ISBN: 0-87664-758-1

Book Design by Summit Technical Associates, Inc.
Cover Design by Lynne Srba Graphics

Dedication

To my dear wife Evelyn, for so many reasons

Preface

I first became directly involved in boiler operation just 50 years ago, in the fall of 1936. I had just started my work toward a mechanical engineering degree at Kansas State University and obtained a live-in part time job in a laundry and dry cleaning shop. My duties included building a coal fire in a hand-fired vertical firetube boiler and having 80 psig steam pressure when the laundry and dry cleaning people came to work at 8:00 AM. Since graduation in 1941, my career has been in the boiler control field, first with the Bailey Meter Co. (now Bailey Controls Co.) and after retirement in early 1980 as an independent consultant.

When I agreed several years ago to write this text, I was not prepared for the effort that would be involved or the length of time before it was finished. That it was completed is one of my life's minor miracles. Along the way boiler and control technology advanced and this required redoing much of the writing that had previously been completed. Other problems were that I am not a natural born writer (though now better than when I started this text) and sometimes went for over a month trying to think of a way to explain a particular aspect of the subject.

But, by far, my major problem was to define the scope of the text and organize the coverage of the subject so that it would be manageable. A basic decision was that the coverage must be limited in scope in order to treat what was covered in at least minor detail. This led to the elimination of the digital logic aspects of boiler control, that of burner and equipment startup and shutdown and the flame safety shutdown actions. Limiting the scope also led to the decision to omit the complex areas of utility boiler control. Both of the above subjects are important enough in their own context that they should be covered in significant detail. This would not have been possible if included here.

In the organization of the book, care was taken to cover the boiler process aspects before the control coverage of those aspects. An attempt was also made to organize the coverage of the control aspects so that they would finally flow together into the complete boiler control system. In the diagrams of complete systems, space limitations required showing a previously shown subsystem as a block of control logic that had previously been covered.

I am very grateful to the many people who helped me along the way and hope that this text will help others to get a running start at boiler control application. I particularly thank E. G. Bailey, the founder of the Bailey Meter Co., for starting the company that provided me with a career that I thoroughly enjoyed. I am happy that I had the opportunity to thank Mr. Bailey personally in 1973 when he was 93 years old and had just received two new patents.

The material in this book is a distillation of my work in the field of boiler control. I thank everyone — hundreds, maybe thousands — who had a part in my training and development; I can honestly say that I learned from everyone. I learned much from talking to operating people who could observe how boilers acted but didn't know why. I learned from other service engineers for boilers, pumps, fans, turbines, switchgear, etc., as we would chat over dinner after the day's testing work. I learned from instrument engineers who worked for consulting engineers as they explained to me how other suppliers handled a particular problem. I learned from control maintenance engineers, design engineers, and operations engineers of client companies with whom I had many discussions concerning control performance and what was needed to improve it. I learned from my competitors in the field as they explained to me why they thought their system was better than mine and I would explain to them why this could not be so. Last, and certainly not least, I learned from my colleagues at Bailey Meter Co. and Babcock and Wilcox Co. who, in the one-way exchange of my early experience and in our normal give and take, passed on their insights and experiences as grist for my mill.

I hope that someone will pick up where I left off and write a text that covers utility boiler control in some detail. Similarly, I hope that someone will write a text that includes in some detail the subject of burner and safety control. Included would be burner management and the startup and shutdown of pulverizers. A third text covering the subject of energy management is needed. Such a text would cover controls for the balancing of industrial heat cycles for least cost operation and also would cover HVAC applications.

For those who read this and are already working in the boiler control field, I hope you will find some insights here that will help you with your work. For those who read this and are working in other fields, I hope that after reading this you will see that there is no mystique and that boiler control is very logical and not a lot of mumbo jumbo.

And now we have distributed digital control. The cost constraints to implementing complex boiler control applications have been virtually eliminated. What a great time for the boiler control application engineer.

Sam Dukelow
Hutchinson, KA

Contents

List of Figures

List of Tables

Section 1
Introduction

1-1 Content and Objectives

This text deals with the control of boilers. In the context of the coverage included, the emphasis will be on high pressure (above 15 psi) steam boilers as applied for industrial power generation and process heat supply. Most of the included material will apply equally to electric utility boilers used for power generation and to smaller boilers for light industrial, commercial and institutional heating applications.

Boiler control is a broad subject that includes the starting and stopping of equipment and the total start-up and shutdown procedures, as well as safety interlocks and the on-line operation of the boiler. This text concentrates on the on-line aspects of boiler control. Boiler control also has a degree of complexity and sophistication that relates generally to the size and complexity of the boiler equipment being controlled. In view of this, control systems for large electric utility boilers are generally more complex than this coverage, which is approximately up to that of the largest industrial steam boiler. The additional complexity of utility boiler control systems is a subject for a more advanced text. Burner control, burner management, start-up and shutdown of equipment, and safety interlocks are also subjects for another text.

The main objective of this text is to introduce on-line boiler control concepts and to develop typical applications to illustrate the use of these concepts. Another objective is the inclusion of the necessary background material so that the reader can properly apply the concepts to his own particular needs.

The text is aimed at those individuals who are actively involved in the operation, engineering, or the sale of boilers and their peripheral equipment, and operation, engineering, sale, or application of boiler control equipment. A knowledge of boiler jargon is therefore assumed. Also assumed is a working knowledge of the thermodynamics relating to boiling, heat, heat transfer, and combustion of fossil fuels. A rudimentary knowledge of control concepts is required and familiarity with the various types of control loops and their tuning characteristics is desirable. The mathematical prerequisite is a secure understanding of arithmetic and algebra.

In order to properly apply control equipment to boilers or any other process, it is necessary to understand the basic aspects of the process that relate to control, the interrelationships of the process characteristics, and the dynamics that are involved. To help the reader develop that understanding, a significant portion of the text discusses the boiler "steaming" process. Another significant portion on boiler fuels and the fuel-burning equipment, their characteristics, and their handling has been included to form the background information for the control of different types of fuels.

1-2 Boiler Control Objectives

For proper control application, it is necessary to understand the objectives of the control system. In the case of steam boilers, there are two basic objectives.

(1) To cause the boiler to provide a continuous supply of steam at the desired condition of pressure and temperature.

(2) To continuously operate the boiler at the lowest cost for fuel and other boiler inputs consistent with high level level of safety and full boiler design life.

The second objective translates into "improving boiler efficiency," since achieving the lowest fuel cost involves operation with the most efficient combustion. For the proper understanding of combustion efficiency and how it is achieved, the text includes material covering the combustion process. This material includes discussion of the measurements that are used to determine combustion and boiler efficiency and the techniques and methods used in determining those efficiency values.

There are a multitude of designs of boiler systems. Built into the designs may be heat recovery features that enable operation at a particular level of cost for fuel and other inputs. Since the automatic control system actually operates the boiler, whether or not the boiler achieves its economy potential is a function of the boiler control system.

Generally, control systems of greater sophistication can control more precisely and come closer to meeting all of the system design objectives; but greater sophistication of a control system usually means a higher initial cost.

It is necessary when applying boiler control systems to understand the trade-offs between increased cost for control sophistication (including a higher level of maintenance) and the savings that result from its application. Investment in control sophistication, as for other investments, usually is layered. The law of "diminishing returns" for each added layer also usually holds true. Each control sophistication improvement should therefore be reviewed on an incremental basis of return relative to investment. To facilitate such analysis, the text develops the control methods by starting with basic control loops and demonstrating added sophistication by optional additions to the base system.

1-3 Control System Diagramming

A boiler control system is an interconnected package of control loops and functions into which a number of inputs are connected and a number of outputs are delivered to final control devices. A change from one input will usually affect more than one output. In addition, a change in one output may have an effect on more than one boiler measurement or input. Because of this, the specific arrangement of the control equipment has a very significant effect on control interaction.

It is a goal in the improvement of this type of control system to minimize these interactions. This requires the development of control logic that will not only perform the control functions but will also minimize the interaction between control loops. To perform these logic funtions, all the basic control functions, feedback (closed-loop), feedforward (open-loop), cascade, and ratio, are used individually and linked together in any needed combination.

This text deals with the logic involved in the control systems and is independent of the type of, or manufacturer of, the control hardware that is used to implement the control schemes. Because of this the SAMA system of control diagramming is used.

While there is an ISA standard for diagramming control systems such as those for boiler control, the ISA system is somewhat hardware-oriented and procedures for showing pure application logic are not as clear-cut as those of the SAMA system. The SAMA system, since it deals only with the control logic involved, is applicable to the older pneumatic or electric analog control, the mechanical control of the James Watt period, and equally to the newer microprocessor control. In addition, the SAMA method has by use become the generally accepted method for diagramming boiler control systems.

In order that all users of this text have a basis for understanding the control diagrams, the essence of the SAMA and ISA systems is given in Table 1-1 and Table 1-2. In addition, Figure 1-1 is a comparative demonstration of the SAMA and ISA diagramming systems.

1-4 Boiler Control Application in Historical Perspective

The inventor of boiler control appears to have been James Watt. Very shortly after he applied the "flyball" governor for speed control of the first rotative steam engines, he applied

Table 1-1
Scientific Apparatus Makers Association Control Diagramming System

TABLE A
ENCLOSURE SYMBOLS

FUNCTION	SYMBOL
MEASURING OR READOUT	◯
MANUAL SIGNAL PROCESSING	◇
AUTOMATIC SIGNAL PROCESSING	▭
FINAL CONTROLLING	⏢

WITHIN A CIRCLE USE A LETTER SYMBOL   FROM TABLE B
WITHIN OTHER ENCLOSURES USE A SYMBOL FROM TABLE C

TABLE B
MEASURING/READOUT LETTERS

PROCESS VARIABLE		FUNCTION
A = ANALYSIS**	**R**	= RECORDING (RECORDER)
C = CONDUCTIVITY		
D = DENSITY	**I**	= INDICATING (INDICATOR)
F = FLOW		
L = LEVEL (FR)	**Q**	= INTEGRATING (TOTALIZER)
M = MOISTURE		
P = PRESSURE (FRT)	**U**	= DIGITAL ACQ. SYSTEM (D.A.S.)
S = SPEED		
T = TEMPERATURE	**T**	= TRANSMITTER
V = VISCOSITY	**RT**	= RECORDING TRANSMITTER
W = WEIGHT	**IT**	= INDICATING TRANSMITTER
Z = POSITION		

**SELF-DEFINING SYMBOLS SUCH AS O_2, pH, ETC., CAN BE USED IN PLACE OF "A".

SYMBOLS

A COMPLETE SYMBOL CONSISTS OF AN ENCLOSURE SYMBOL (TABLE A) WITHIN WHICH IS CONTAINED A MEASURING OR READOUT LETTER (TABLE B), OR A SIGNAL PROCESSING SYMBOL (TABLE C). COMPLEX FUNCTIONS ARE REPRESENTED BY THE COMBINATIONS OF THESE BASIC SYMBOLS.

TABLE C
SIGNAL PROCESSING SYMBOLS

FUNCTION	SIGNAL PROCESSING SYMBOL	FUNCTION	SIGNAL PROCESSING SYMBOL
SUMMING	Σ or $+$	INTEGRATE OR TOTALIZE	**Q**
AVERAGING	Σ/n	HIGH SELECTING	$>$
DIFFERENCE	$\triangle$ or $-$	LOW SELECTING	$<$
PROPORTIONAL	**K** or **P**	HIGH LIMITING	⊅
INTEGRAL	$\int$ or **I**	LOW LIMITING	⊄
DERIVATIVE	d/dt or **D**	REVERSE PROPORTIONAL	$-$**K** or $-$**P**
MULTIPLYING	**X**	VELOCITY LIMITING	**V**⊅
DIVIDING	$\div$	BIAS	$\pm$
ROOT EXTRACTION	$\sqrt[n]{\ }$	TIME FUNCTION	**f(t)**
EXPONENTIAL	x^n	VARIABLE SIGNAL GENERATOR	**A**
NON $-$ LINEAR FUNCTION	**f(x)**	TRANSFER	**T**
TRI $-$ STATE SIGNAL (RAISE, HOLD, LOWER)	$\updownarrow$	SIGNAL MONITOR	**H/, H/L, /L**

Table 1-2
ISA Control Diagramming System

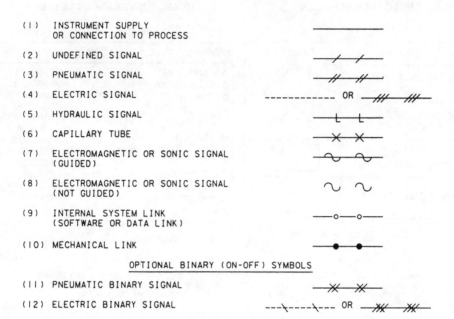

INSTRUMENT LINE SYMBOLS

(1) INSTRUMENT SUPPLY
        OR CONNECTION TO PROCESS

(2) UNDEFINED SIGNAL

(3) PNEUMATIC SIGNAL

(4) ELECTRIC SIGNAL

(5) HYDRAULIC SIGNAL

(6) CAPILLARY TUBE

(7) ELECTROMAGNETIC OR SONIC SIGNAL
        (GUIDED)

(8) ELECTROMAGNETIC OR SONIC SIGNAL
        (NOT GUIDED)

(9) INTERNAL SYSTEM LINK
        (SOFTWARE OR DATA LINK)

(10) MECHANICAL LINK

OPTIONAL BINARY (ON-OFF) SYMBOLS

(11) PNEUMATIC BINARY SIGNAL

(12) ELECTRIC BINARY SIGNAL

GENERAL INSTRUMENT OR FUNCTION SYMBOLS

	PRIMARY LOCATION NORMALLY ACCESSIBLE TO OPERATOR	FIELD MOUNTED	AUXILIARY LOCATION NORMALLY ACCESSIBLE TO OPERATOR
DISCRETE INSTRUMENTS	1	2	3
SHARED DISPLAY, SHARED CONTROL	4	5	6
COMPUTER FUNCTION	7	8	9
PROGRAMMABLE LOGIC CONTROL	10	11	12

Table 1-2
(Continued)

NO	FUNCTION	SYMBOL
1	SUMMING	Σ
2	AVERAGING	Σ/n
3	DIFFERENCE	Δ
4	PROPORTIONAL	K 1:1 2:1
5	INTEGRAL	$\int$
6	DERIVATIVE	d/dt
7	MULTIPLYING	X
8	DIVIDING	$\div$
9	ROOT EXTRACTION	$\sqrt[n]{}$

NO	FUNCTION	SYMBOL
10	EXPONENTIAL	x^n
11	NONLINEAR OR UNSPECIFIED FUNCTION	$f(X)$
12	TIME FUNCTION	$f(t)$
13	HIGH SELECTING	$>$
14	LOW SELECTING	$<$
15	HIGH LIMITING	$\not{>}$
16	LOW LIMITING	$\not{<}$
17	REVERSE PROPORTIONAL	-K
18	VELOCITY LIMITER	$\forall$

Table 1-2
(continued)

NO	FUNCTION	SYMBOL
19	BIAS	+
		−
		+/−
20	CONVERT	*/*

NO	FUNCTION	SYMBOL
21	SIGNAL MONITOR	**H
		**L
		**HL

IDENTIFICATION LETTERS

	FIRST-LETTER (4)		SUCCEEDING-LETTERS (3)		
	MEASURED OR INITIATING VARIABLE	MODIFIER	READOUT OR PASSIVE FUNCTION	OUTPUT FUNCTION	MODIFIER
A	Analysis		Alarm		
B	Burner, Combustion		User's Choice	User's Choice	User's Choice
C	User's Choice			Control	
D	User's Choice	Differential			
E	Voltage		Sensor (Primary Element)		
F	Flow Rate	Ratio (Fraction)			
G	User's Choice		Glass, Viewing Device		
H	Hand				High
I	Current (Electrical)		Indicate		
J	Power	Scan			
K	Time, Time Schedule	Time Rate of Change		Control Station	
L	Level		Light		Low
M	User's Choice	Momentary			Middle, Intermediate
N	User's Choice		User's Choice	User's Choice	User's Choice
O	User's Choice		Orifice, Restriction		
P	Pressure, Vacuum		Point (Test) Connection		
Q	Quantity	Integrate, Totalize			
R	Radiation		Record		
S	Speed, Frequency	Safety		Switch	

Table 1-2
(continued)

IDENTIFICATION LETTERS

	FIRST-LETTER (4)		SUCCEEDING-LETTERS (3)		
	MEASURED OR INITIATING VARIABLE	MODIFIER	READOUT OR PASSIVE FUNCTION	OUTPUT FUNCTION	MODIFIER
T	Temperature			Transmit	
U	Multivariable		Multifunction	Multifunction	Multifunction
V	Vibration, Mechanical Analysis			Valve, Damper, Louver	
W	Weight, Force		Well		
X	Unclassified	X Axis	Unclassified	Unclassified	Unclassified
Y	Event, State or Presence	Y Axis		Relay, Compute, Convert	
Z	Position, Dimension	Z Axis		Driver, Actuator, Unclassified Final Control Element	

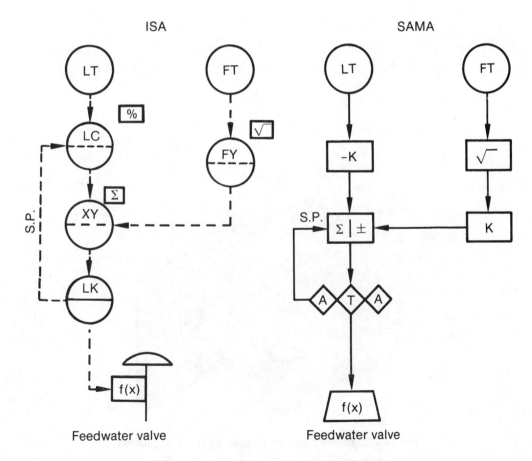

Figure 1-1 Two-Element Feedwater Control

feedback control to automatically control the level in the boiler drum by regulating the water to the boiler. Within approximately 10 years of that time he also applied feedback control to automatically control steam pressure by regulating the boiler draft. This is shown in Figure 1-2, a copy of a drawing of a boiler from that time period.

From that time in the late 1790's, while there were some improvements in the hardware used, the application concepts of boiler control did not advance until the early 20th century. From approximately 1915 until 1950, boiler control developed into integrated systems for combustion control, feedwater control, and steam temperature control. This period also covered the acceptance phase of this type of equipment. By 1950 boiler combustion control had proven its worth, and it was accepted that any new boiler installation would include the installation of automatic boiler control equipment.

In the period between 1950 and 1970, while there were very significant developments in control application for utility boilers and the most complex large industrial boilers, the development of industrial boiler control was primarily hardware-oriented. During this period there was a progressive increase in the use of the concept of implementing boiler control by linking together analog computing devices. A second hardware-oriented development for new installations during this period was a switch from predominantly pneumatic analog control to predominantly electronic analog control.

On the negative side, industrial boiler control regressed during this 1950 to 1970 period, due to the continual reduction in "constant dollar" fuel prices relative to the cost of boilers and their appurtenances. The result was the use of less control sophistication for the average new installation. As this situation progressed, larger and larger boiler installations, with their

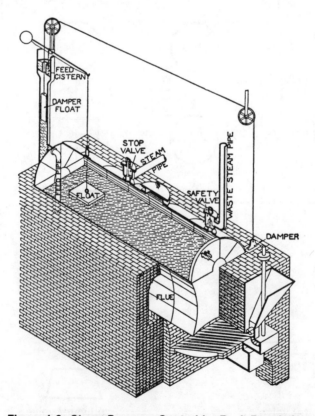

Figure 1-2. Steam Pressure Control by Draft Regulation

(circa 1790) (From *Types of Boilers*, Carl S. Dow, S.B. 1911) increased consumption of fuel, were required in order to economically justify the more complex boiler control systems. Figure 1-3 demonstrates this with a comparison between the cost of fuel oil and the cost of boiler control systems of comparable complexity.

Since 1970 the economic balance has completely turned around (see Figure 1-3). The very high price of fuel in the '80's can justify on any boiler a much greater degree of control sophistication than could be justified in 1970. In addition, the development of microprocessor control has sparked a beneficial transition to the greater precision of digital control. The development of new sensors has been instrumental in the development of new boiler control application concepts. This text should be viewed in the context of the equipment (both measurement and control) that exists today and the fuel economics that have driven the changes of recent years.

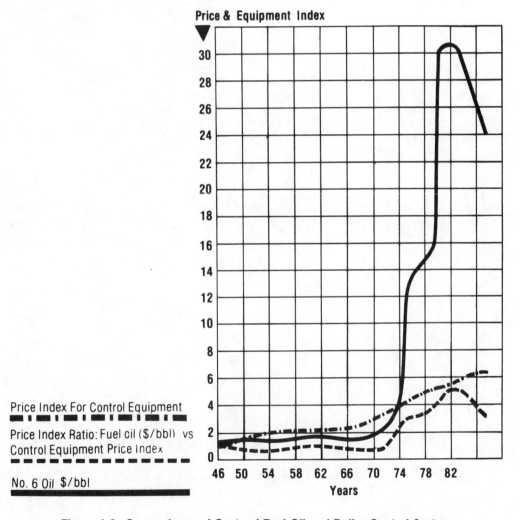

Figure 1-3. Comparisons of Costs of Fuel Oil and Boiler Control Systems

Section 2
Boiler Basics and the Steaming Process

2-1 The Basic Steaming Process

In the conversion of water from its liquid phase to steam, its vapor phase, heat is added to initially increase the water temperature to the boiling point temperature. This heat that raises the temperature of the water is known as sensible heat. The boiling point temperature is 212° F at atmospheric pressure and rises as the pressure in the system is increased. The boiling point temperature is also known as the saturation temperature of the steam that is produced. The relationships between the saturation temperatures and pressures of steam are fixed thermodynamic properties of steam.

As the conversion from the liquid phase (water) to the vapor phase (steam) begins, the temperature no longer changes with the addition of heat. The fluid exists at the saturation temperature-pressure relationship during the entire conversion of the water to steam. The heat that is added in converting from the liquid to the vapor phase at constant temperature is called the latent heat of evaporation. Steam that is not fully vaporized is called "wet" steam. The percentage by weight of the water droplets in the wet steam is known as the % moisture. The % quality of wet steam is obtained by subtracting the % moisture from 100.

The total amount of heat in a quantity of steam includes the amount of sensible heat above 32° F and the latent heat of evaporation. Generally, as the pressure of dry saturated steam (100 percent conversion to vapor and at saturation temperature) increases, the amount of sensible heat increases and the amount of latent heat decreases. The relationships between the various thermodynamic steam properties are shown in Tables 2-1 and 2-2.

By adding additional sensible heat to dry saturated steam, the temperature can be increased above the saturation temperature. Steam that is heated above the saturation temperature is called superheated steam. The effect on the thermodynamic properties by superheating steam is shown in Table 2-3. Note that superheating increases the total heat or enthalphy (h) of the steam (h = Btu/lb; Btu = heat required to raise 1 lb. of water from 59° to 60°). Superheating also causes the steam to expand, increasing its specific volume (cu ft/lb).

2-2 The Basic Boiler

A basic diagram of a boiler is shown in Figure 2-1. This diagram shows that there are two separate systems in a boiler. One system is the steam-water system, which is also called the water side of the boiler. Into this system water is introduced and, upon receiving heat that is transferred through a solid metal barrier, is heated, converted to steam, and leaves the system in the form of steam.

The other system of a boiler is the fuel-air-flue gas system which is also called the fire side of the boiler. This system provides the heat that is transferred to the water. The inputs to this system are fuel and the necessary air required to burn the fuel.

In this system the fuel and air are thoroughly mixed and ignited in a furnace. The resulting combustion converts the chemical energy of the fuel to thermal or heat energy. The furnace is usually lined with heat transfer surface in the form of water-steam circulating tubes. These tubes receive heat radiating from the flame and transfer it to the water-side system. The gases resulting from the combustion, known as the flue gases, are cooled by the transfer of their heat by what is known as the radiant heat transfer surface. The gases leave the furnace and pass through additional heating surface that is in the form of water-steam circulating tubes. In this area the surfaces cannot "see" the flame, and the heat is transferred by convection. This area is known as the convection heating surface. In this section addi-

Table 2-1
Saturation: Temperatures

Temp F t	Abs Press. Lb Sq In. p	Specific Volume Sat. Liquid v_f	Evap. v_{fg}	Sat. Vapor v_g	Enthalpy Sat. Liquid h_f	Evap. h_{fg}	Sat. Vapor h_g	Entropy Sat. Liquid s_f	Evap. s_{fg}	Sat. Vapor s_g	Temp F t
32	0.08854	0.01602	3306	3306	0.00	1075.8	1075.8	0.0000	2.1877	2.1877	32
35	0.09995	0.01602	2947	2947	3.02	1074.1	1077.1	0.0061	2.1709	2.1770	35
40	0.12170	0.01602	2444	2444	8.05	1071.3	1079.3	0.0162	2.1435	2.1597	40
45	0.14752	0.01602	2036.4	2036.4	13.06	1068.4	1081.5	0.0262	2.1167	2.1429	45
50	0.17811	0.01603	1703.2	1703.2	18.07	1065.6	1083.7	0.0361	2.0903	2.1264	50
60	0.2563	0.01604	1206.6	1206.7	28.06	1059.9	1088.0	0.0555	2.0393	2.0948	60
70	0.3631	0.01606	867.8	867.9	38.04	1054.3	1092.3	0.0745	1.9902	2.0647	70
80	0.5069	0.01608	633.1	633.1	48.02	1048.6	1096.6	0.0932	1.9428	2.0360	80
90	0.6982	0.01610	468.0	468.0	57.99	1042.9	1100.9	0.1115	1.8972	2.0087	90
100	0.9492	0.01613	350.3	350.4	67.97	1037.2	1105.2	0.1295	1.8531	1.9826	100
110	1.2748	0.01617	265.3	265.4	77.94	1031.6	1109.5	0.1471	1.8106	1.9577	110
120	1.6924	0.01620	203.25	203.27	87.92	1025.8	1113.7	0.1645	1.7694	1.9339	120
130	2.2225	0.01625	157.32	157.34	97.90	1020.0	1117.9	0.1816	1.7296	1.9112	130
140	2.8886	0.01629	122.99	123.01	107.89	1014.1	1122.0	0.1984	1.6910	1.8894	140
150	3.718	0.01634	97.06	97.07	117.89	1008.2	1126.1	0.2149	1.6537	1.8685	150
160	4.741	0.01639	77.27	77.29	127.89	1002.3	1130.2	0.2311	1.6174	1.8485	160
170	5.992	0.01645	62.04	62.06	137.90	996.3	1134.2	0.2472	1.5822	1.8293	170
180	7.510	0.01651	50.21	50.23	147.92	990.2	1138.1	0.2630	1.5480	1.8109	180
190	9.339	0.01657	40.94	40.96	157.95	984.1	1142.0	0.2785	1.5147	1.7932	190
200	11.526	0.01663	33.62	33.64	167.99	977.9	1145.9	0.2938	1.4824	1.7762	200
210	14.123	0.01670	27.80	27.82	178.05	971.6	1149.7	0.3090	1.4508	1.7598	210
212	14.696	0.01672	26.78	26.80	180.07	970.3	1150.4	0.3120	1.4446	1.7566	212
220	17.186	0.01677	23.13	23.15	188.13	965.2	1153.4	0.3239	1.4201	1.7440	220
230	20.780	0.01684	19.365	19.382	198.23	958.8	1157.0	0.3387	1.3901	1.7288	230
240	24.969	0.01692	16.306	16.323	208.34	952.2	1160.5	0.3531	1.3609	1.7140	240
250	29.825	0.01700	13.804	13.821	218.48	945.5	1164.0	0.3675	1.3323	1.6998	250
260	35.429	0.01709	11.746	11.763	228.64	938.7	1167.3	0.3817	1.3043	1.6860	260
270	41.858	0.01717	10.044	10.061	238.84	931.8	1170.6	0.3958	1.2769	1.6727	270
280	49.203	0.01726	8.628	8.645	249.06	924.7	1173.8	0.4096	1.2501	1.6597	280
290	57.556	0.01735	7.444	7.461	259.31	917.5	1176.8	0.4234	1.2238	1.6472	290
300	67.013	0.01745	6.449	6.466	269.59	910.1	1179.7	0.4369	1.1980	1.6350	300
320	89.66	0.01765	4.896	4.914	290.28	894.9	1185.2	0.4637	1.1478	1.6115	320
340	118.01	0.01787	3.770	3.788	311.13	879.0	1190.1	0.4900	1.0992	1.5891	340
360	153.04	0.01811	2.939	2.957	332.18	862.2	1194.4	0.5158	1.0519	1.5677	360
380	195.77	0.01836	2.317	2.335	353.45	844.6	1198.1	0.5413	1.0059	1.5471	380
400	247.31	0.01864	1.8447	1.8633	374.97	826.0	1201.0	0.5664	0.9608	1.5272	400
420	308.83	0.01894	1.4811	1.5000	396.77	806.3	1203.1	0.5912	0.9166	1.5078	420
440	381.59	0.01926	1.1979	1.2171	418.90	785.4	1204.3	0.6158	0.8730	1.4887	440
460	466.9	0.0196	0.9748	0.9944	441.4	763.2	1204.6	0.6402	0.8298	1.4700	460
480	566.1	0.0200	0.7972	0.8172	464.4	739.4	1203.7	0.6645	0.7868	1.4513	480
500	680.8	0.0204	0.6545	0.6749	487.8	713.9	1201.7	0.6887	0.7438	1.4325	500
520	812.4	0.0209	0.5385	0.5594	511.9	686.4	1198.2	0.7130	0.7006	1.4136	520
540	962.5	0.0215	0.4434	0.4649	536.6	656.6	1193.2	0.7374	0.6568	1.3942	540
560	1133.1	0.0221	0.3647	0.3868	562.2	624.2	1186.4	0.7621	0.6121	1.3742	560
580	1325.8	0.0228	0.2989	0.3217	588.9	588.4	1177.3	0.7872	0.5659	1.3532	580
600	1542.9	0.0236	0.2432	0.2668	617.0	548.5	1165.5	0.8131	0.5176	1.3307	600
620	1786.6	0.0247	0.1955	0.2201	646.7	503.6	1150.3	0.8398	0.4664	1.3062	620
640	2059.7	0.0260	0.1538	0.1798	678.6	452.0	1130.5	0.8679	0.4110	1.2789	640
660	2365.4	0.0278	0.1165	0.1442	714.2	390.2	1104.4	0.8987	0.3485	1.2472	660
680	2708.1	0.0305	0.0810	0.1115	757.3	309.9	1067.2	0.9351	0.2719	1.2071	680
700	3093.7	0.0369	0.0392	0.0761	823.3	172.1	995.4	0.9905	0.1484	1.1389	700
705.4	3206.2	0.0503	0	0.0503	902.7	0	902.7	1.0580	0	1.0580	705.4

tional amounts of heat are transferred to the water side of the boiler. This heat transfer further cools the flue gases which then leave the boiler.

Since heat transfer depends upon a temperature difference as a "driving force", with the simple boiler described the flue gases can be cooled only to a temperature that is at some level above the temperature of the steam-water system. The temperature of the flue gases determines the amount of heat remaining in these gases, so the heat loss in the boiler flue gases is determined to some extent by the saturation temperature in the steam-water system.

The process of adding heat to convert water to steam has a time constant that depends upon the specific characteristics of the installation. The factors affecting this time constant include the system heat storage, the heat transfer coefficients in different parts of the system,

Table 2-2
Saturation: Pressures

Abs Press. Lb Sq In. P	Temp F t	Specific Volume Sat. Liquid v_f	Specific Volume Sat. Vapor v_g	Enthalpy Sat. Liquid h_f	Enthalpy Evap h_{fg}	Enthalpy Sat. Vapor h_g	Entropy Sat. Liquid s_f	Entropy Evap s_{fg}	Entropy Sat. Vapor s_g	Internal Energy Sat. Liquid u_f	Internal Energy Evap u_{fg}	Internal Energy Sat. Vapor u_g	Abs Press. Lb Sq In. P
1.0	101.74	0.01614	333.6	69.70	1036.3	1106.0	0.1326	1.8456	1.9782	69.70	974.6	1044.3	1.0
2.0	126.08	0.01623	173.73	93.99	1022.2	1116.2	0.1749	1.7451	1.9200	93.98	957.9	1051.9	2.0
3.0	141.48	0.01630	118.71	109.37	1013.2	1122.6	0.2008	1.6855	1.8863	109.36	947.3	1056.7	3.0
4.0	152.97	0.01636	90.63	120.86	1006.4	1127.3	0.2198	1.6427	1.8625	120.85	939.3	1060.2	4.0
5.0	162.24	0.01640	73.52	130.13	1001.0	1131.1	0.2347	1.6094	1.8441	130.12	933.0	1063.1	5.0
6.0	170.06	0.01645	61.98	137.96	996.2	1134.2	0.2472	1.5820	1.8292	137.94	927.5	1065.4	6.0
7.0	176.85	0.01649	53.64	144.76	992.1	1136.9	0.2581	1.5586	1.8167	144.74	922.7	1067.4	7.0
8.0	182.86	0.01653	47.34	150.79	988.5	1139.3	0.2674	1.5383	1.8057	150.77	918.4	1069.2	8.0
9.0	188.28	0.01656	42.40	156.22	985.2	1141.4	0.2759	1.5203	1.7962	156.19	914.6	1070.8	9.0
10	193.21	0.01659	38.42	161.17	982.1	1143.3	0.2835	1.5041	1.7876	161.14	911.1	1072.2	10
14.696	212.00	0.01672	26.80	180.07	970.3	1150.4	0.3120	1.4446	1.7566	180.02	897.5	1077.5	14.696
15	213.03	0.01672	26.29	181.11	969.7	1150.8	0.3135	1.4415	1.7549	181.06	896.7	1077.8	15
20	227.96	0.01683	20.089	196.16	960.1	1156.3	0.3356	1.3962	1.7319	196.10	885.8	1081.9	20
30	250.33	0.01701	13.746	218.82	945.3	1164.1	0.3680	1.3313	1.6993	218.73	869.1	1087.8	30
40	267.25	0.01715	10.498	236.03	933.7	1169.7	0.3919	1.2844	1.6763	235.90	856.1	1092.0	40
50	281.01	0.01727	8.515	250.09	924.0	1174.1	0.4110	1.2474	1.6585	249.93	845.4	1095.3	50
60	292.71	0.01738	7.175	262.09	915.5	1177.6	0.4270	1.2168	1.6438	261.90	836.0	1097.9	60
70	302.92	0.01748	6.206	272.61	907.9	1180.6	0.4409	1.1906	1.6315	272.38	827.8	1100.2	70
80	312.03	0.01757	5.472	282.02	901.1	1183.1	0.4531	1.1676	1.6207	281.76	820.3	1102.1	80
90	320.27	0.01766	4.896	290.56	894.7	1185.3	0.4641	1.1471	1.6112	290.27	813.4	1103.7	90
100	327.81	0.01774	4.432	298.40	888.8	1187.2	0.4740	1.1286	1.6026	298.08	807.1	1105.2	100
120	341.25	0.01789	3.728	312.44	877.9	1190.4	0.4916	1.0962	1.5878	312.05	795.6	1107.6	120
140	353.02	0.01802	3.220	324.82	868.2	1193.0	0.5069	1.0682	1.5751	324.35	785.2	1109.6	140
160	363.53	0.01815	2.834	335.93	859.2	1195.1	0.5204	1.0436	1.5640	335.39	775.8	1111.2	160
180	373.06	0.01827	2.532	346.03	850.8	1196.9	0.5325	1.0217	1.5542	345.42	767.1	1112.5	180
200	381.79	0.01839	2.288	355.36	843.0	1198.4	0.5435	1.0018	1.5453	354.68	759.0	1113.7	200
250	400.95	0.01865	1.8438	376.00	825.1	1201.1	0.5675	0.9588	1.5263	375.14	740.7	1115.8	250
300	417.33	0.01890	1.5433	393.84	809.0	1202.8	0.5879	0.9225	1.5104	392.79	724.3	1117.1	300
350	431.72	0.01913	1.3260	409.69	794.2	1203.9	0.6056	0.8910	1.4966	408.45	709.6	1118.0	350
400	444.59	0.0193	1.1613	424.0	780.5	1204.5	0.6214	0.8630	1.4844	422.6	695.9	1118.5	400
450	456.28	0.0195	1.0320	437.2	767.4	1204.6	0.6356	0.8378	1.4734	435.5	683.2	1118.7	450
500	467.01	0.0197	0.9278	449.4	755.0	1204.4	0.6487	0.8147	1.4634	447.6	671.0	1118.6	500
550	476.93	0.0199	0.8422	460.8	743.1	1203.9	0.6608	0.7934	1.4542	458.8	659.2	1118.2	550
600	486.21	0.0201	0.7698	471.6	731.6	1203.2	0.6720	0.7734	1.4454	469.4	648.3	1117.7	600
700	503.10	0.0205	0.6554	491.5	709.7	1201.2	0.6925	0.7371	1.4296	488.8	627.5	1116.3	700
800	518.23	0.0209	0.5687	509.7	688.9	1198.6	0.7108	0.7045	1.4153	506.6	607.8	1114.4	800
900	531.98	0.0212	0.5006	526.6	668.8	1195.4	0.7275	0.6744	1.4020	523.1	589.0	1112.1	900
1000	544.61	0.0216	0.4456	542.4	649.4	1191.8	0.7430	0.6467	1.3897	538.4	571.0	1109.4	1000
1100	556.31	0.0220	0.4001	557.4	630.4	1187.8	0.7575	0.6205	1.3780	552.9	553.5	1106.4	1100
1200	567.22	0.0223	0.3619	571.7	611.7	1183.4	0.7711	0.5956	1.3667	566.7	536.3	1103.0	1200
1300	577.46	0.0227	0.3293	585.4	593.2	1178.6	0.7840	0.5719	1.3559	580.0	519.4	1099.4	1300
1400	587.10	0.0231	0.3012	598.7	574.7	1173.4	0.7963	0.5491	1.3454	592.7	502.7	1095.4	1400
1500	596.23	0.0235	0.2765	611.6	556.3	1167.9	0.8082	0.5269	1.3351	605.1	486.1	1091.2	1500
2000	635.82	0.0257	0.1878	671.7	463.4	1135.1	0.8619	0.4230	1.2849	662.2	403.4	1065.6	2000
2500	668.13	0.0287	0.1307	730.6	360.5	1091.1	0.9126	0.3197	1.2322	717.3	313.3	1030.6	2500
3000	695.36	0.0346	0.0858	802.5	217.8	1020.3	0.9731	0.1885	1.1615	783.4	189.3	972.7	3000
3206.2	705.40	0.0503	0.0503	902.7	0	902.7	1.0580	0	1.0580	872.9	0	872.9	3206.2

the masses of metal and refractory and their configuration, and various other factors. For the purpose of control, it is generally enough to understand that the complete time constant is a matter of minutes. Viewing the system as achieving 63 per cent of total response in one fifth of the total time constant is sufficient for most boiler control analysis procedures.

2-3 Heat Recovery from the Flue Gases

If the heat losses in the boiler flue gases are to be reduced, separate heat exchangers must be added to the simple boiler to recover more of the heat and further cool the flue gases. The combustion air preheater is one form of such an added heat exchanger. The application of an air preheater is shown in Figure 2-2. The flue gas leaves the boiler and passes through the combustion air preheater. The combustion air also passes through the air preheater before being mixed with the fuel. Since the flue gas temperature is higher than the air temperature, heat is transferred from the flue gas to the combustion air via the convection heat transfer surface of the combustion air preheater.

Table 2-3
Superheated Vapor

Abs Press. Lb/Sq In. (Sat. Temp)		200	300	400	500	600	700	800	900	1000	1200	1400	1600
1 (101.74)	v	392.6	452.3	512.0	571.6	631.2	690.8	750.4	809.9	869.5	988.7	1107.8	1227.0
	h	1150.4	1195.8	1241.7	1288.3	1335.7	1383.8	1432.8	1482.7	1533.5	1637.7	1745.7	1857.5
	s	2.0512	2.1153	2.1720	2.2233	2.2702	2.3137	2.3542	2.3923	2.4283	2.4952	2.5566	2.6137
5 (162.24)	v	78.16	90.25	102.26	114.22	126.16	138.10	150.03	161.95	173.87	197.71	221.6	245.4
	h	1148.8	1195.0	1241.2	1288.0	1335.4	1383.6	1432.7	1482.6	1533.4	1637.7	1745.7	1857.4
	s	1.8718	1.9370	1.9942	2.0456	2.0927	2.1361	2.1767	2.2148	2.2509	2.3178	2.3792	2.4363
10 (193.21)	v	38.85	45.00	51.04	57.05	63.03	69.01	74.98	80.95	86.92	98.84	110.77	122.69
	h	1146.6	1193.9	1240.6	1287.5	1335.1	1383.4	1432.5	1482.4	1533.2	1637.6	1745.6	1857.3
	s	1.7927	1.8595	1.9172	1.9689	2.0160	2.0596	2.1002	2.1383	2.1744	2.2413	2.3028	2.3598
14.696 (212.00)	v		30.53	34.68	38.78	42.86	46.94	51.00	55.07	59.13	67.25	75.37	83.48
	h		1192.8	1239.9	1287.1	1334.8	1383.2	1432.3	1482.3	1533.1	1637.5	1745.5	1857.3
	s		1.8160	1.8743	1.9261	1.9734	2.0170	2.0576	2.0958	2.1319	2.1989	2.2603	2.3174
20 (227.96)	v		22.36	25.43	28.46	31.47	34.47	37.46	40.45	43.44	49.41	55.37	61.34
	h		1191.6	1239.2	1286.6	1334.4	1382.9	1432.1	1482.1	1533.0	1637.4	1745.4	1857.2
	s		1.7808	1.8396	1.8918	1.9392	1.9829	2.0235	2.0618	2.0978	2.1648	2.2263	2.2834
40 (267.25)	v		11.040	12.628	14.168	15.688	17.198	18.702	20.20	21.70	24.69	27.68	30.66
	h		1186.8	1236.5	1284.8	1333.1	1381.9	1431.3	1481.4	1532.4	1637.0	1745.1	1857.0
	s		1.6994	1.7608	1.8140	1.8619	1.9058	1.9467	1.9850	2.0212	2.0883	2.1498	2.2069
60 (292.71)	v		7.259	8.357	9.403	10.427	11.441	12.449	13.452	14.454	16.451	18.446	20.44
	h		1181.6	1233.6	1283.0	1331.8	1380.9	1430.5	1480.8	1531.9	1636.6	1744.8	1856.7
	s		1.6492	1.7135	1.7678	1.8162	1.8605	1.9015	1.9400	1.9762	2.0434	2.1049	2.1621
80 (312.03)	v			6.220	7.020	7.797	8.562	9.322	10.077	10.830	12.332	13.830	15.325
	h			1230.7	1281.1	1330.5	1379.9	1429.7	1480.1	1531.3	1636.2	1744.5	1856.5
	s			1.6791	1.7346	1.7836	1.8281	1.8694	1.9079	1.9442	2.0115	2.0731	2.1303
100 (327.81)	v			4.937	5.589	6.218	6.835	7.446	8.052	8.656	9.860	11.060	12.258
	h			1227.6	1279.1	1329.1	1378.9	1428.9	1479.5	1530.8	1635.7	1744.2	1856.2
	s			1.6518	1.7085	1.7581	1.8029	1.8443	1.8829	1.9193	1.9867	2.0484	2.1056
120 (341.25)	v			4.081	4.636	5.165	5.683	6.195	6.702	7.207	8.212	9.214	10.213
	h			1224.4	1277.2	1327.7	1377.8	1428.1	1478.8	1530.2	1635.3	1743.9	1856.0
	s			1.6287	1.6869	1.7370	1.7822	1.8237	1.8625	1.8990	1.9664	2.0281	2.0854
140 (353.02)	v			3.468	3.954	4.413	4.861	5.301	5.738	6.172	7.035	7.895	8.752
	h			1221.1	1275.2	1326.4	1376.8	1427.3	1478.2	1529.7	1634.9	1743.5	1855.7
	s			1.6087	1.6683	1.7190	1.7645	1.8063	1.8451	1.8817	1.9493	2.0110	2.0683
160 (363.53)	v			3.008	3.443	3.849	4.244	4.631	5.015	5.396	6.152	6.906	7.656
	h			1217.6	1273.1	1325.0	1375.7	1426.4	1477.5	1529.1	1634.5	1743.2	1855.5
	s			1.5908	1.6519	1.7033	1.7491	1.7911	1.8301	1.8667	1.9344	1.9962	2.0535
180 (373.06)	v			2.649	3.044	3.411	3.764	4.110	4.452	4.792	5.466	6.136	6.804
	h			1214.0	1271.0	1323.5	1374.7	1425.6	1476.8	1528.6	1634.1	1742.9	1855.2
	s			1.5745	1.6373	1.6894	1.7355	1.7776	1.8167	1.8534	1.9212	1.9831	2.0404
200 (381.79)	v			2.361	2.726	3.060	3.380	3.693	4.002	4.309	4.917	5.521	6.123
	h			1210.3	1268.9	1322.1	1373.6	1424.8	1476.2	1528.0	1633.7	1742.6	1855.0
	s			1.5594	1.6240	1.6767	1.7232	1.7655	1.8048	1.8415	1.9094	1.9713	2.0287
220 (389.86)	v			2.125	2.465	2.772	3.066	3.352	3.634	3.913	4.467	5.017	5.565
	h			1206.5	1266.7	1320.7	1372.6	1424.0	1475.5	1527.5	1633.3	1742.3	1854.7
	s			1.5453	1.6117	1.6652	1.7120	1.7545	1.7939	1.8308	1.8987	1.9607	2.0181
240 (397.37)	v			1.9276	2.247	2.533	2.804	3.068	3.327	3.584	4.093	4.597	5.100
	h			1202.5	1264.5	1319.2	1371.5	1423.2	1474.8	1526.9	1632.9	1742.0	1854.5
	s			1.5319	1.6003	1.6546	1.7017	1.7444	1.7839	1.8209	1.8889	1.9510	2.0084

v—Specific Volume (cu. ft./lb.)
h—Total Heat (BTU/lb.)
s—Entropy

Table 2-3
(continued)

Abs Press. Lb/Sq In. (Sat. Temp)		500	600	700	800	900	1000	1200	1400	1600
260 (404.42)	v	2.063	2.330	2.582	2.827	3.067	3.305	3.776	4.242	4.707
	h	1262.3	1317.7	1370.4	1422.3	1474.2	1526.3	1632.5	1741.7	1854.2
	s	1.5897	1.6447	1.6922	1.7352	1.7748	1.8118	1.8799	1.9420	1.9995
280 (411.05)	v	1.9047	2.156	2.392	2.621	2.845	3.066	3.504	3.938	4.370
	h	1260.0	1316.2	1369.4	1421.5	1473.5	1525.8	1632.1	1741.4	1854.0
	s	1.5796	1.6354	1.6834	1.7265	1.7662	1.8033	1.8716	1.9337	1.9912
300 (417.33)	v	1.7675	2.005	2.227	2.442	2.652	2.859	3.269	3.674	4.078
	h	1257.6	1314.7	1368.3	1420.6	1472.8	1525.2	1631.7	1741.0	1853.7
	s	1.5701	1.6268	1.6751	1.7184	1.7582	1.7954	1.8638	1.9260	1.9835
350 (431.72)	v	1.4923	1.7036	1.8980	2.084	2.266	2.445	2.798	3.147	3.493
	h	1251.5	1310.9	1365.5	1418.5	1471.1	1523.8	1630.7	1740.3	1853.1
	s	1.5481	1.6070	1.6563	1.7002	1.7403	1.7777	1.8463	1.9086	1.9663
400 (444.59)	v	1.2851	1.4770	1.6508	1.8161	1.9767	2.134	2.445	2.751	3.055
	h	1245.1	1306.9	1362.7	1416.4	1469.4	1522.4	1629.6	1739.5	1852.5
	s	1.5281	1.5894	1.6398	1.6842	1.7247	1.7623	1.8311	1.8936	1.9513
450 (456.28)	v	1.1231	1.3005	1.4584	1.6074	1.7516	1.8928	2.170	2.443	2.714
	h	1238.4	1302.8	1359.9	1414.3	1467.7	1521.0	1628.6	1738.7	1851.9
	s	1.5095	1.5735	1.6250	1.6699	1.7108	1.7486	1.8177	1.8803	1.9381
500 (467.01)	v	0.9927	1.1591	1.3044	1.4405	1.5715	1.6996	1.9504	2.197	2.442
	h	1231.3	1298.6	1357.0	1412.1	1466.0	1519.6	1627.6	1737.9	1851.3
	s	1.4919	1.5588	1.6115	1.6571	1.6982	1.7363	1.8056	1.8683	1.9262
550 (476.94)	v	0.8852	1.0431	1.1783	1.3038	1.4241	1.5414	1.7706	1.9957	2.219
	h	1223.7	1294.3	1354.0	1409.9	1464.3	1518.2	1626.6	1737.1	1850.6
	s	1.4751	1.5451	1.5991	1.6452	1.6868	1.7250	1.7946	1.8575	1.9155
600 (486.21)	v	0.7947	0.9463	1.0732	1.1899	1.3013	1.4096	1.6208	1.8279	2.033
	h	1215.7	1289.9	1351.1	1407.7	1462.5	1516.7	1625.5	1736.3	1850.0
	s	1.4586	1.5323	1.5875	1.6343	1.6762	1.7147	1.7846	1.8476	1.9056
700 (503.10)	v		0.7934	0.9077	1.0108	1.1082	1.2024	1.3853	1.5641	1.7405
	h		1280.6	1345.0	1403.2	1459.0	1513.9	1623.5	1734.8	1848.8
	s		1.5084	1.5665	1.6147	1.6573	1.6963	1.7666	1.8299	1.8881
800 (518.23)	v		0.6779	0.7833	0.8763	0.9633	1.0470	1.2088	1.3662	1.5214
	h		1270.7	1338.6	1398.6	1455.4	1511.0	1621.4	1733.2	1847.5
	s		1.4863	1.5476	1.5972	1.6407	1.6801	1.7510	1.8146	1.8729
900 (531.98)	v		0.5873	0.6863	0.7716	0.8506	0.9262	1.0714	1.2124	1.3509
	h		1260.1	1332.1	1393.9	1451.8	1508.1	1619.3	1731.6	1846.3
	s		1.4653	1.5303	1.5814	1.6257	1.6656	1.7371	1.8009	1.8595
1000 (544.61)	v		0.5140	0.6084	0.6878	0.7604	0.8294	0.9615	1.0893	1.2146
	h		1248.8	1325.3	1389.2	1448.2	1505.1	1617.3	1730.0	1845.0
	s		1.4450	1.5141	1.5670	1.6121	1.6525	1.7245	1.7886	1.8474
1100 (556.31)	v		0.4532	0.5445	0.6191	0.6866	0.7503	0.8716	0.9885	1.1031
	h		1236.7	1318.3	1384.3	1444.5	1502.2	1615.2	1728.4	1843.8
	s		1.4251	1.4989	1.5535	1.5995	1.6405	1.7130	1.7775	1.8363
1200 (567.22)	v		0.4016	0.4909	0.5617	0.6250	0.6843	0.7967	0.9046	1.0101
	h		1223.5	1311.0	1379.3	1440.7	1499.2	1613.1	1726.9	1842.5
	s		1.4052	1.4843	1.5409	1.5879	1.6293	1.7025	1.7672	1.8263
1400 (587.10)	v		0.3174	0.4062	0.4714	0.5281	0.5805	0.6789	0.7727	0.8640
	h		1193.0	1295.5	1369.1	1433.1	1493.2	1608.9	1723.7	1840.0
	s		1.3639	1.4567	1.5177	1.5666	1.6093	1.6836	1.7489	1.8083

(Tables 2-1, 2-2, and 2-3 are from *Steam, Its Generation and Use*, © Babcock and Wilcox.)

This transfer of heat cools the flue gas and thus reduces its heat loss. The added heat in the combustion air enters the furnace, enhances the combustion process, and reduces the fuel requirement in an amount equal in heat value to the heat that has been transferred in the combustion air preheater. By the use of an air preheater, approximately 1 per cent of fuel is saved for each 40° F rise in the combustion air temperature.

The use of an economizer is another flue gas heat recovery method. The arrangement of this type of additional heat exchanger is shown in Figure 2-3. In the economizer arrangement shown, the flue gas leaves the simple boiler and enters the economizer, where it is in contact with heat transfer surface in the form of water tubes through which the boiler feedwater flows. Since the flue gas is at a higher temperature than the water, the flue gas is cooled and the water temperature is increased. Cooling the flue gas reduces its heat loss in an amount equal to the increased heat in the feedwater to the boiler. The increased heat in the feedwater reduces the boiler's requirement for fuel and combustion air. Approximately 1 percent of fuel input is saved for each 10° F rise in the feedwater as it passes through the economizer.

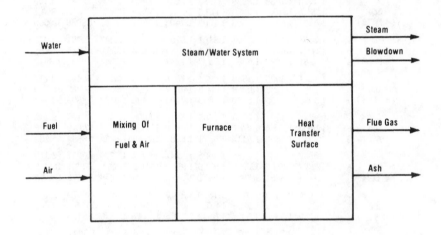

Figure 2-1 Basic Diagram of a Boiler

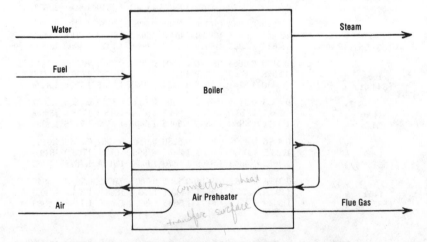

Air Preheater Purpose—Preheat combustion air and absorb additional heat from flue gases

Figure 2-2 A Simple Boiler plus Combustion Air Preheater

Both types of heat exchangers are often used in large boilers. When both an air preheater and an economizer are used, the normal practice consists of passing the flue gases first through the economizer and then through the combustion air preheater.

2-4 Boiler Types and Classifications

There are two general types of boilers: firetube and watertube. In addition, boilers are classified as "high" or "low" pressure and as "steam" boilers or "hot water" boilers.

By definition high pressure boilers are steam boilers that operate at a pressure greater than 15 psig. Because the boiler water temperature rises as the pressure is increased, the flue gas temperature is increased as the pressure increases, increasing the boiler heat losses.

An advantage of using higher pressure is a reduction in physical size of the boiler and steam piping for the same heat-carrying capacity. This is due to the increased density (lower specific volume) of the higher pressure steam. The advantage is particularly important if the boiler is some distance from the heat load. When high pressure boilers are used for space heating, the pressure is usually reduced near the point of steam use.

A particular attribute of high pressure steam is that it contains a significantly greater amount of available energy. Available energy is a term given to the energy that is available to be converted to work in an industrial or electric power generation steam engine or turbine.

A low pressure boiler is one that is operated at a pressure lower than 15 psig. Almost all low pressure boilers are used for space heating. Low pressure boiler systems are simpler since pressure reducing valves are seldom required and the water chemistry of the boiler is simpler to maintain.

Another boiler classification is the hot water boiler. Strictly speaking, this is not a boiler since the water does not boil. It is essentially a fuel fired hot water heater in which sensible heat is added to increase the temperature to some level below the boiling point. Because of similarities in many ways to steam boilers, the term "hot water boiler" is generally used to describe this type of unit.

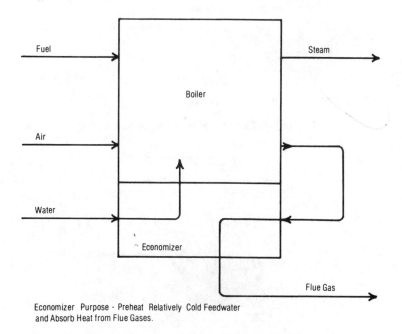

Economizer Purpose - Preheat Relatively Cold Feedwater
and Absorb Heat from Flue Gases.

Figure 2-3 A Simple Boiler plus Economizer

A high temperature hot water (HTHW) boiler furnishes water at a temperature greater than 250° (121°C) or at a prssure higher than 160 psig. A low temperature hot water boiler furnishes water at a pressure not exceeding 160 psig and at a temperature not exceeding 250° F.

2-5 Firetube Boilers

Firetube boilers constitute the largest share of small- to medium-sized industrial units. In firetube boilers the flue gas products of combustion flow through boiler tubes surrounded by water. Steam is generated by the heat transferred through the walls of the tubes to the surrounding water. The flue gases are cooled as they flow through the tubes, transferring their heat to the water; therefore, the cooler the flue gas, the greater the amount of heat transferred. Cooling of the flue gas is a function of the heat conductivity of the tube and its surfaces, the temperature difference between the flue gases and the water in the boiler, the heat transfer area, the time of contact between the flue gases and the boiler tube surface, and other factors.

Firetube boilers used today evolved from the earliest designs of a spherical or cylindrical pressure vessel mounted over the fire with flame and hot gases around the boiler shell. This obsolete approach has been improved by installing longitudinal tubes in the pressure vessel and passing flue gases through the tubes. This increased the heat transfer area and improved the heat transfer coefficient. The results are the two variations of the horizontal return tubular (HRT) boiler shown in Figures 2-4 and 2-5. A variation of the HRT boiler in Figure 2-4 is the packaged (shop-assembled) firebox boiler shown in Figure 2-6.

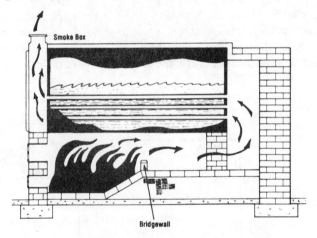

Figure 2-4 Horizontal-Return-Tubular Boiler

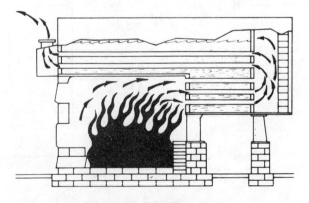

Figure 2-5 Two-Pass Boiler

A parallel evolution of the firetube boiler was the locomotive boiler designed with the furnace surrounded by a heat transfer area and a heat transfer area added by using horizontal tubes. This type is shown in Figure 2-7.

The Scotch Marine boiler design, as shown in Figure 2-8 with the furnace a large metal tube, combined that feature of the English Cornish boiler of the 1800's and the smaller horizontal tubes of the HRT boiler. This boiler originally was developed to fit the need for compact shipboard boilers. Because the furnace is cooled completely by water, no refractory furnace is required. The radiant heat from the combustion is transferred directly through the metal wall of the furnace chamber to the water. This allows the furnace walls to become a heat transfer surface — a surface particularly effective because of the high temperature differential between the flame and the boiler water.

A modified Scotch boiler design, as used in the standard firetube package boiler, is the most common firetube boiler used today. There are two variations of the Scotch design. These are called wetback and dryback and are shown in Figures 2-9 and 2-10. These names refer to the rear of the combustion chamber, which must be either water-jacketed or lined with a high temperature insulating material, such as refractory, to protect it from the heat of combustion.

The wetback boiler gains some additional heating surface; however, it is more difficult to service because access to the back end of the boiler tubes is limited. The only such access normally provided is a 16-inch manhole in the rear water header or through the furnace tube.

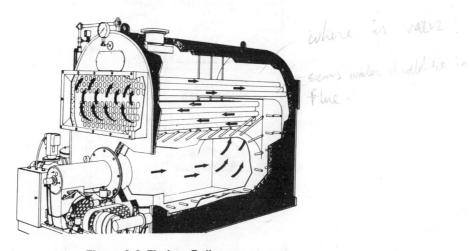

where is water?
seems water should be in flue.

Figure 2-6 Firebox Boiler

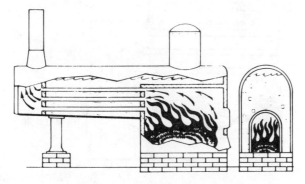

Figure 2-7 Locomotive-type Boiler

The dryback boiler is easy to service because the rear doors may be removed for complete access to the tubes and to the insulating or refractory material. The refractory or insulating lining may deteriorate over a period of time. If this lining is not properly maintained, efficiency may be reduced because the flue gases will bypass heating surface on three- and four-pass designs, the radiation loss through the rear doors will increase, and the metal doors will be damaged.

The number of boiler passes for a firetube boiler refers to the number of horizontal runs the flue gases take between the furnace and the flue gas outlet. The combustion chamber or

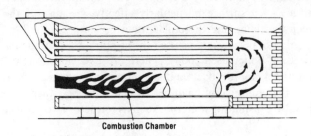

Combustion Chamber

Figure 2-8 Scotch Marine Boiler

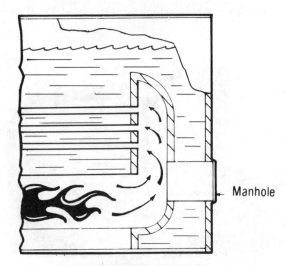

Manhole

Figure 2-9 Wetback

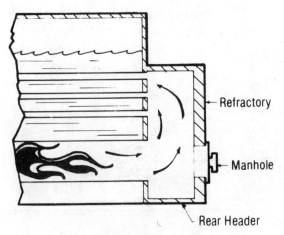

Refractory

Manhole

Rear Header

Figure 2-10 Dryback

furnace is considered the first pass; each separate set of firetubes provides additional passes as shown in Figure 2-11.

The number of gas passes in a firetube boiler does not necessarily determine its efficiency characteristic. For the same total number, length, and size of tubes (same tube heating surface), increasing the number of passes increases the length the flue gas must travel because the gases must pass through tubes in series rather than in parallel. This increases the flue gas velocity within the tubes but does little to change the total time for the hot gases to flow from furnace to outlet in contact with the tube heating surface.

The increased gas velocity in some cases may improve heat transfer by increasing the turbulence of the gases as they travel through the tubes. Generally, however, increasing the number of passes and the resultant velocity of the gases increases the resistance to flow and forces the combustion air blower to consume more power.

One additional firetube boiler, generally used only where space is limited and steam requirements are small, is the vertical firetube boiler shown in Figure 2-12. This is a variation of the firebox boiler with the water-jacketed furnace and vertical tubes.

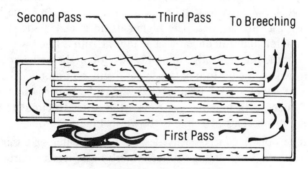

Figure 2-11 Boiler Passes

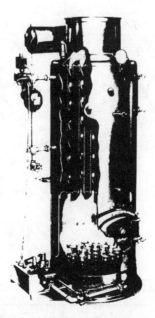

Figure 2-12 Vertical Firetube Boiler

Characteristics of the various types of firetube boilers relative to operational limitations are approximate in Table 2-4.

Table 2-4
Firetube Boiler Characteristics (Approximate)

Boiler Type	Max Pressure	BoHP* Range	Lbs/Hr
HRT	150 psig	30–300	1000–10000
Firebox	200 psig	10–600	350–25000
Pkg. "Scotch"	300 psig	10–1000	350–35000
Vert Firetube	200 psig	2–300	70–10000

* The term BoHP is discussed in Section 3

2-6 Watertube Boilers

As the name implies, water circulates within the tubes of a watertube boiler. These tubes are usually connected between two or more cylindrical drums. In some boilers the lower drum is replaced with a tube header. The higher drum is called the steam drum and is maintained approximately half full of water. The lower drum is filled with water completely and is the low point of the boiler. Sludge that may develop in the boiler gravitates to the low point and can be drawn off the bottom of this lower drum, commonly called the mud drum.

A cross-sectional view of a small field erected watertube boiler is shown in Figure 2-13. Heating the "riser" tubes with hot flue gas causes the water to circulate and steam to be released in the steam drum. This principle is shown in Figure 2-14.

Because watertube boilers can be easily designed for greater or lesser furnace volume using the same boiler convection heating surface, watertube boilers are particularly applicable to solid fuel firing. They are also applicable for a full range of sizes and for pressures from 50 psig to 5000 psig. The present readily available minimum size of industrial watertube

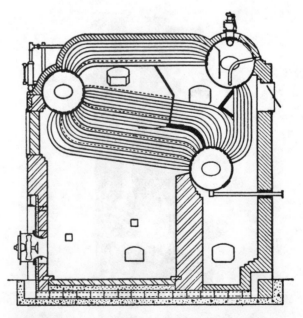

Figure 2-13 Small Field-Erected Watertube Boiler

boilers is approximately 20,000 to 25,000 lbs/hr of steam — equivalent to 600 to 750 BoHP (boiler horsepower). Many watertube boilers operating today are in the 250 to 300 BoHP size range.

The maximum size of electric utility watertube boilers, at this writing, handles approximately 10,000,000 lbs/hr of steam. In industrial use the largest are in the approximate range of 1,000,000 to 1,500,000 lbs/hr.

A typical industrial watertube boiler for gas and oil firing is the packaged (shop-assembled) boiler shown in Figure 2-15. Such packaged watertube boilers generally have a single burner up to approximately 125,000 lbs/hr steam flow (approximately 4000 BoHP) and are furnished in sizes up to approximately 250,000 lbs/hr.

Older designs of watertube boilers, as shown in Figure 2-13, consisted of refractory-lined furnaces and with only convection heating surface. Later developments included placing

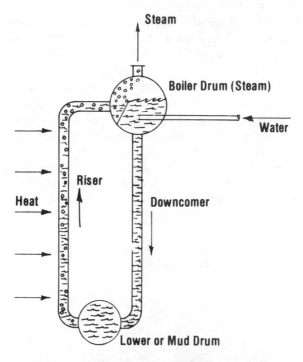

Figure 2-14 Circulation of Watertube Boiler

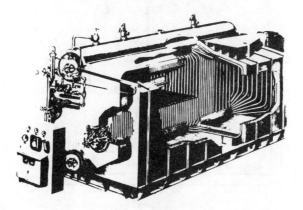

Figure 2-15 Packaged Watertube Boiler

some boiler tubes in the furnace walls where they were exposed to the radiant heat of the flame as shown in Figure 2-16. This development continued until furnaces became fully water-walled, as in the package boiler shown in Figure 2-15.

This evolution in the design of boilers came about because of the economic benefits resulting from the lower cost of the required radiant heat transfer surface as compared with convection heating surface. The net result was a reduction in the physical size and cost of the boilers and changes in water volume, heat storage, and an improvement in response characteristics. Other effects of the fully water-cooled furnaces were reduced furnace temperatures and the resulting reduction in NOx production. The cooler furnaces also affect the chemistry of the progression of the combustion process from ignition to complete combustion.

As industrial boilers are increased in size for liquid and gaseous fuels, the balance between radiant and convection heat transfer surface remains approximately the same as for the package boiler for oil and gas firing shown in Figure 2-15. One such larger gas- or oil-fired industrial boiler is shown in in Figure 2-17.

For solid fuel, however, coal, wood, or waste material-fired boilers usually require greater spacing between the boiler tubes. In addition the furnace volume must be increased. A large industrial boiler for solid fuel is shown in Figure 2-18. These differences make it difficult to convert a gas- or oil-fired boiler to coal and obtain full steam capacity, while a conversion from coal to gas or oil firing can be much more easily accomplished.

Boilers used in electric utility generating plants are generally larger than their industrial counterparts. The steam generating capacity of the largest electric utility boiler is approximately 10 times the capacity of the largest industrial steam boiler. Modern utility boilers operate at pressures in the range of 2,000 to 4,000 psig, while their industrial counterparts are generally in the range of 100 to 1000 psig. In generating electric power with a turbogenerator, it is much more efficient to use steam that has been superheated and reheated as is done in the typical electric utility plant. The general practice with industrial boilers is to use saturated steam or only small amounts of superheat unless electric power is being generated in the industrial plant.

A modern electric utility boiler, as shown in Figure 2-19, may have only the steam drum. Because the water used is very pure, chemical sludge does not normally develop and that

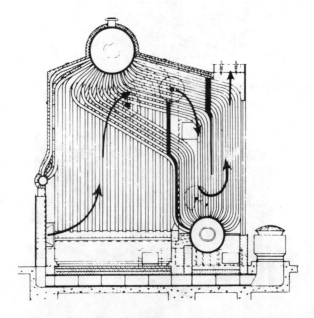

Figure 2-16 Flue Gas Flow and Watertube Boiler

need for the mud drum is eliminated. The lower end of the circulation loop consists of water headers. Boilers such as shown can be designed to operate up to approximately 2,750 psig. For higher operating pressures, the boiler is a "once-through" design as shown in Figure 2-20. In this design, water is pumped into one end of the boiler tubes and superheated steam emerges. This type of boiler would be very rarely if ever used for industrial steam.

Figure 2-16 shows the baffles for directing flue gas. Watertube boilers generally have such gas baffles to assure contact between the hot gases and the maximum amount of the tube heating surface. The baffle design determines the number of gas passes and which tubes act as "risers" and "downcomers" as shown in Figure 2-14. Leakage in the baffles causes hot flue gas to bypass a portion of the heating surface, thereby decreasing the heat being transferred and lowering the boiler efficiency.

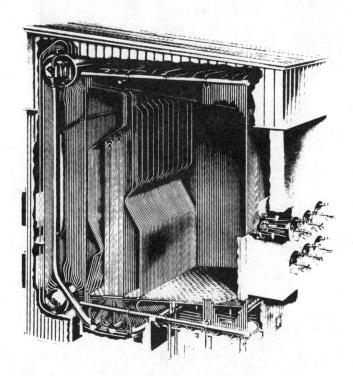

Figure 2-17 Gas- or Oil-Fired Industrial Boiler

(From *Steam, Its Generation and Use*, © Babcock and Wilcox. Used with permission.)

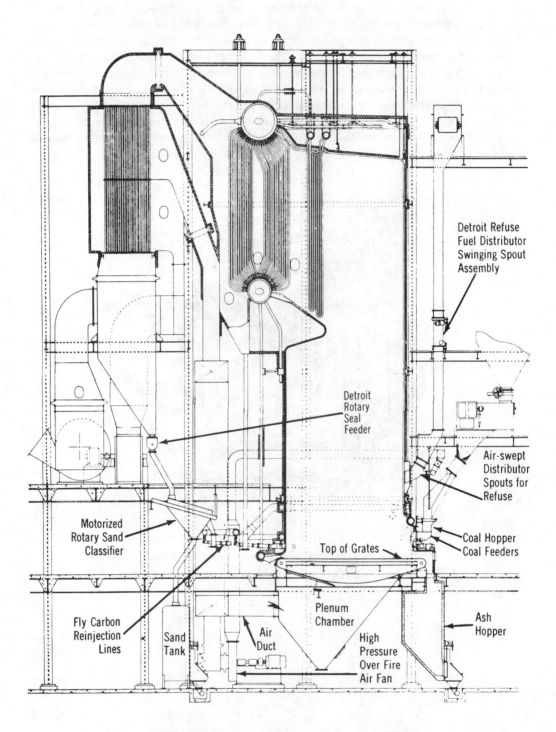

Detroit Refuse
Fuel Distributor
Swinging Spout
Assembly

Detroit
Rotary
Seal
Feeder

Air-swept
Distributor
Spouts for
Refuse

Motorized
Rotary Sand
Classifier

Top of Grates

Coal Hopper
Coal Feeders

Fly Carbon
Reinjection
Lines

Sand
Tank

Air
Duct

Plenum
Chamber

High
Pressure
Over Fire
Air Fan

Ash
Hopper

Figure 2-18 Large Industrial Boiler for Solid Fuel

(From Detroit Stoker Company. Used with permission.)

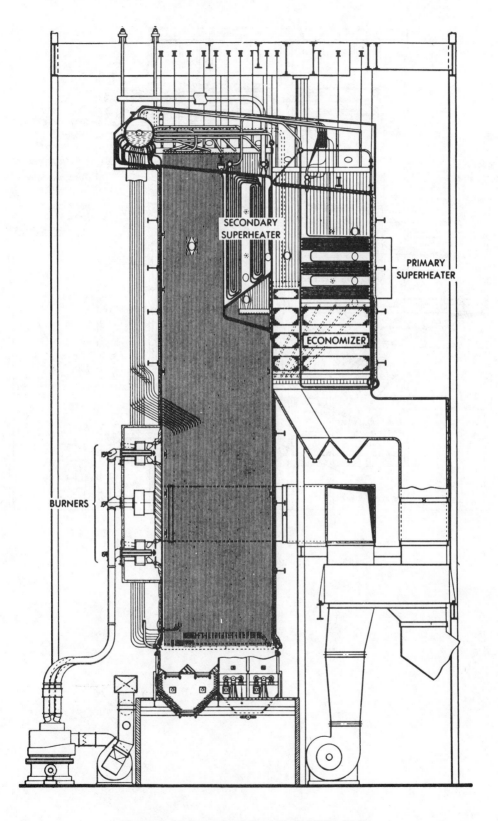

Figure 2-19 Pulverized Coal-Fired Utility Boiler

(From *Steam, Its Geration and Use*, © Babcock and Wilcox)

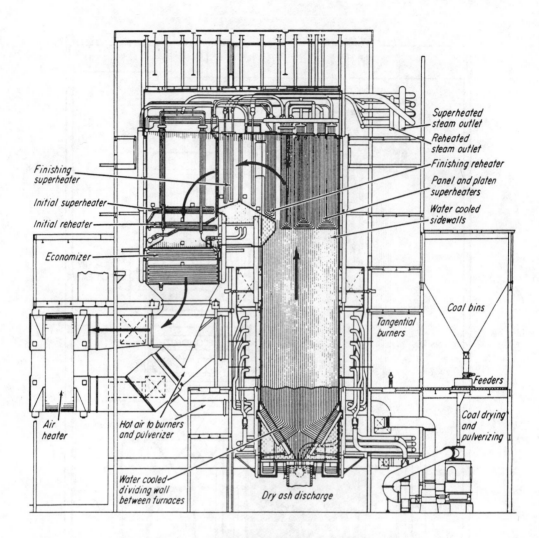

Figure 2-20 Large 3500-psig Combined Circulation Boiler

(Courtesy Combustion Engineering Corporation)

Section 3
Performance and Input/Output Relationships

A boiler's performance relates to its ability to transfer heat from the fuel to the water while meeting operating specifications. Boiler performance includes all aspects of the operation. The basic elements are the operating capacity and the boiler efficiency.

Performance specifications include the operating capacity and the factors for adjusting that capacity, steam pressure, boiler water quality, boiler temperatures, boiler pressures, boiler drafts and draft losses, flue gas analysis, fuel analysis, and fuel burned. Additional performance specifications indicate the fan power requirements (boiler flue gas temperatures and draft losses) and the fuel supply assumptions.

The result of a calculation involving the performance specification is a calculated efficiency. Boiler efficiency is presented as a percentage ratio of heat supplied to the boiler and the heat absorbed in the boiler water.

3-1 Capacity and Performance

Packaged firetube boilers generally are described in terms of BoHP (boiler horsepower).* The BoHP rating of a modern firetube boiler is approximately one-fifth of its square feet of heating surface. For example, a boiler of 500 BoHP has approximately 2,500 square feet of heating surface. Although these boilers are described in terms of BoHP, the developed Btu output can be converted easily to lbs/hr of steam. Because the heat content of a pound of steam increases as pressure is increased in firetube boilers, the pounds of steam per BoHP decreases with pressure. Table 3-1 shows this relationship.

Table 3-1
Conversion Factors — Lbs Steam per BoHP

Boiler Pressure, psig	Lbs Steam/BoHP	Btu Content from 212° F (liquid)
50	33.8	999
75	33.6	1005.2
100	33.4	1009.6
125	33.31	1013
150	33.22	1015.6
175	33.16	1017.6
200	33.1	1019.3
225	33.06	1020.6
250	33.03	1021.7

---

*The term "boiler horsepower" started as a result of early boilers used to drive engines with one engine horsepower or one boiler horsepower equivalent to 34.5 pounds of water evaporated from and at 212 degrees. This equals 33,475 Btu, thermal equivalent of one boiler horsepower.

Industrial watertube boilers formerly were classified in BoHP by dividing the heating surface by 10. In the past 30 or more years, however, BoHP ratings for new watertube boilers have disappeared and boiler capacity ratings are specified in terms of pounds of steam per hour with feedwater temperature specified. Existing watertube boilers rated in BoHP can be rated in lbs/hr by using a conversion factor from Table 3-1. Smaller watertube and firetube boilers often are rated in terms of maximum Btu input to the burner with efficiency specified.

3-2 Input Related to Output

Boiler energy inputs generally are thought of as the heat content of the fuel used. The flow of this fuel measured over a period of time multiplied by the heat content of this fuel develops a total Btu input during the time period. Measuring the energy output of a steam boiler involves measuring the steam flow in lbs/hr over a period of time and multiplying by the Btu content of a pound of steam to provide the Btu output. Useful simple relationships of input and output — such as pounds of steam/gallon of fuel oil, pounds of steam/pound of coal, or pounds of steam/standard cubic foot of gas — can be used to track relative efficiency. These relationships, however, are not precise because such factors as fuel Btu content, steam Btu content, feedwater temperature, and blowdown are not considered.

The chief energy loss of most boilers depends on the mass of the flue gases and their temperature as they leave the boiler. To obtain the net energy loss of the flue gas, however, the temperatures of the incoming combustion air and fuel must be considered.

When hydrogen in the fuel burns, it forms water, which leaves the boiler in the form of superheated vapor. The latent heat of this vapor is an energy loss, which is approximately nine to ten percent for natural gas, five to six percent for fuel oil, and three to four percent for coal. The percentage of hydrogren and moisture in the fuel affects this loss.

Although blowdown is not a useful heat output from the viewpoint of boiler efficiency, it is not considered a loss because the boiler has properly transferred the heat from the fuel to the water.

The useful energy output of boilers is the heat carried by the steam or hot water. In a steam boiler this is usually measured as steam flow at the boiler and adjusted for Btu content by measurements of pressure, temperature, or both. The steam flow can also be obtained by measuring water flow and subtracting the blowdown. For hot water boilers, water flow is measured at the boiler outlet and adjusted for Btu content by measurement of the outlet temperature.

Although these procedures provide information about the useful energy outputs, in themselves they do not determine precisely the contribution of the boiler to this useful energy. To determine the contribution of the boiler, the heat in the incoming feedwater must be subtracted from the heat carried in the boiler output.

3-3 Mass and Energy Balances Involved

The mass balances in a steam boiler are shown in the diagrams of Figures 3-1 through 3-5. For a hot water boiler these diagrams would be slightly different. In Figure 3-1 there is a simple balance on the water side of the boiler between the mass of the feedwater and the mass of the steam plus blowdown. In this balance, steam is normally 90 to 99 percent of the output.

Figure 3-2 represents the balance between the mass of combustion air plus fuel and the flue gas and ash output. Ash, of course, would not be present if there were no ash in the fuel. The combustion air is by far the larger input because it may have a mass of more than twelve to eighteen times that of the fuel.

Chemical input and output on the boiler water side also must be considered as one of the mass balances involved. This is shown in Figure 3-3. In this case there is a mass balance of each individual chemical element present. Steam is expected to be so pure that almost 100 percent of the non-water chemical output is in the boiler blowdown.

The balance of chemical input and output of the combustion process is shown in Figure 3-4. As with the water side chemical balance, this diagram represents a balance of each chemical element although the chemical compounds of the inputs have been changed to different chemical compounds by the combustion process.

The energy balance of the boiler is shown in Figure 3-5. Energy enters and leaves a boiler in a variety of ways. Energy in the steam is the only output considered useful. Fuel energy is by far the major energy input and is normally the only energy input considered.

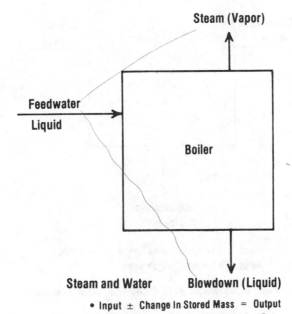

Steam (Vapor)

Feedwater

Liquid

Boiler

Steam and Water   **Blowdown (Liquid)**

- Input ± Change In Stored Mass = Output
- Stored Mass Decreases With Steaming Rate

Figure 3-1 Steam-Water Mass Balance

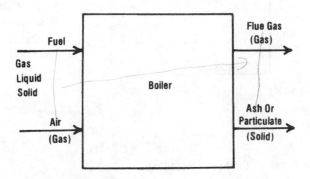

Flue Gas
(Gas)

Fuel

Gas
Liquid
Solid

Boiler

Air

(Gas)

Ash Or
Particulate

(Solid)

Fuel, Air, and Flue Gas
- Input = Output ± Deposits In Boiler.

Figure 3-2 Fuel, Air — Flue Gas Mass Balance

3-4 Efficiency Calculation Methods

Two methods of calculating the efficiency of a boiler are acceptable. These are generally known as the input/output method and the heat loss method. Both methods will be covered in more detail in Section 6, Boiler Efficiency Computations. The input/output method depends on the measurements of fuel, steam, and feedwater flow and the heat content of each.

$$\text{Boiler Efficiency} = \frac{\text{Heat Added to Incoming Feedwater}}{\text{Heat Input (Fuel)}}$$

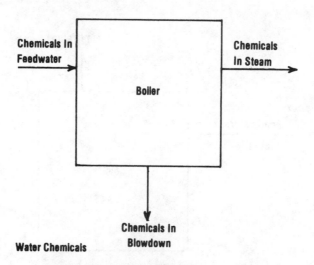

Figure 3-3 Water Side Chemical Mass Balance

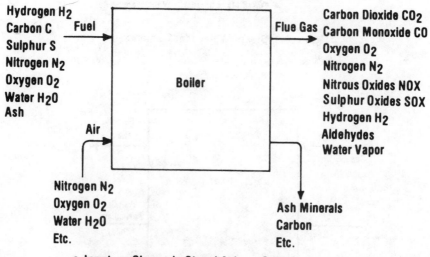

Figure 3-4 Mass Balance — Fuel and Air Chemicals

In this formula, the boiler is credited with the heat added to the blowdown portion of the feedwater. This method yields a decimal number fraction, which is expressed as percent of efficiency. In the heat loss method, the percentage of each of the major losses is determined. To their total, a small percentage for unaccounted loss is added, and the total obtained is subtracted from 100 percent. There are eight major losses:

(1) Sensible heat loss in the dry flue gas
(2) Sensible heat loss from water in the combustion air
(3) Sensible heat loss from water in the fuel
(4) Latent heat loss from water in the fuel
(5) Latent heat loss from water formed by hydrogen combustion
(6) Loss from unburned carbon in the refuse
(7) Loss from unburned combustible gas in the flue gas
(8) Heat loss from radiation

3-5 Boiler Control — The Process of Managing the Energy and Mass Balances

The boiler control system is the vehicle through which the boiler energy and mass balances are managed. All the boiler major energy and mass inputs must be regulated in order to achieve the desired output conditions. The measurements of the output process variables furnish the information to the control system intelligence unit. Figure 3-6 is a block diagram showing how the parts of the overall control system are coordinated into the overall boiler control system.

For the energy input requirement, a firing rate demand signal must be developed. This firing rate demand creates the separate demands for the mass of fuel and combustion air. The mass of the water-steam energy carrier must also be regulated, and the feedwater control

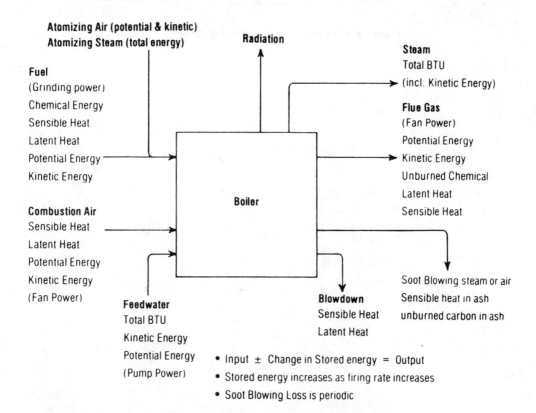

Figure 3-5 Energy Balance — Heat Balance

regulates the mass of water in the boiler. The final steam temperature condition must also be regulated (for boilers generating superheated steam and having such control capability), and this is accomplished by the steam temperature control system. The effects of the input control actions interact, since firing rate also affects steam temperature and feedwater flow affects the steam pressure, which is the final arbiter of firing rate demand. The overall system must therefore be applied and coordinated in a manner to minimize the effect of these interactions.

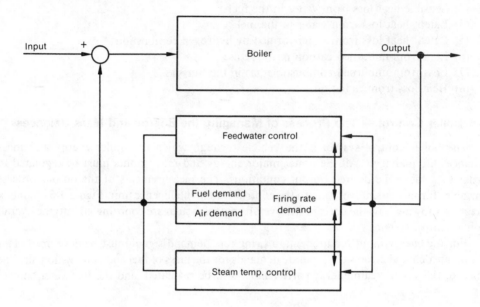

Figure 3-6 Block Diagram of Boiler Control System

Section 4
Basic Control Loops and Their Interconnections

Boiler control systems are normally multivariable with the control loops for fuel, combustion air, and feedwater interacting in the overall system. The overall system generally can be shown to be a series of basic control loops connected together. Boiler control systems can be easily understood if one has a good basic knowledge of these loops and their application requirements.

In the descriptions of the loop, the term primary variable is given to that measurement of a process variable that is to be maintained at a "set point" by the control action. The term manipulated variable is given to that process device that is manipulated in order to achieve the desired condition or set point of the primary variable. The controller function is the brain that determines the magnitude and direction of the changes to the manipulated variable.

4-1 Simple Feedback Control

Simple feedback control is shown in the control diagram of Figure 4-1. With this type of loop, changes in the primary variable feed back to a control function as shown. The control function can be proportional-plus-integral as shown, proportional only, proportional-plus-derivative, integral only, or proportional-plus-integral-plus-derivative. In all these cases the controller includes an error detector function, which measures the error between the primary variable and the set point.

The controller output is determined by the gain (multiplier) and/or time constants that are tuned into the controller funtion device. The controller output changes the manipulated variable, which changes the process output selected as the primary variable, and thus closes

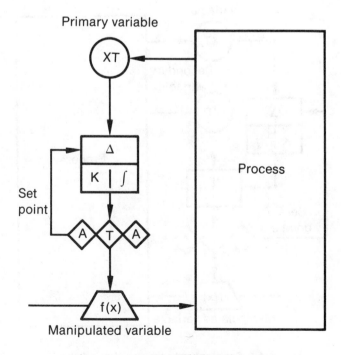

Figure 4-1 Simple Feedback Control

the control loop. There are only two basic types of control, feedforward and feedback. These are used as building blocks in forming all types of modulating control action.

4-2 Feedforward-plus-Feedback Control

In feedforward-plus-feedback control as shown in Figure 4-2, a secondary variable that has a predictable relationship with the manipulated variable is connected. In this case a change in the secondary variable causes the manipulated variable to change in anticipation of a change in the primary variable. This reduces the magnitude of the primary variable change due the more timely control action that originates from the secondary variable. The feedback portion of the loop contains the set point and can contain any of the controller functions of the basic feedback loop.

4-3 Cascade Control

Cascade control consists essentially of two feedback control loops connected together with the output of the primary loop acting as set point for the secondary loop. Cascade control, Figure 4-3, is applied to stabilize the manipulated variable so that a predictable relationship between the manipulated variable and the primary variable can be maintained.

To avoid control instability due to interaction between the two feedback control loops, it is necessary that the response time constants be substantially different, with the process response of the secondary control loop being the faster of the two. A general rule is that the time constant of the primary loop process response should be a minimum of 5 to 10 times that of the secondary loop. The longer time constant of the primary loop indicates a much slower response. Because of this, a normal application would be temperature control (a normally slow loop) cascading onto flow control (a normally fast loop). Other suitable candidates for cascade control are temperature cascading onto pressure and level control cascading onto flow control.

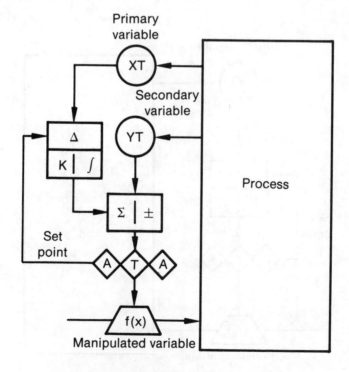

Figure 4-2 Feedforward-plus-Feedback Control

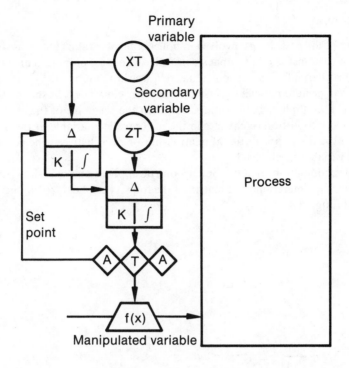

Figure 4-3 Cascade Control

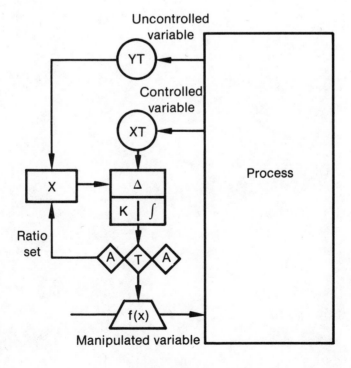

Figure 4-4 Ratio Control

4-4 Ratio Control

The fourth of the basic loops involved in boiler control is ratio control as shown in Figure 4-4. Ratio control consists of a feedback controller whose set point is in direct proportion to an uncontrolled variable. The proportional relationship can be set by the operator of the process, or it can be automatically adjusted by another controller. As shown, the mathematical function is a multiplier. If the ratio is set, then the set point of the controlled variable changes in direct proportion to changes in the uncontrolled variable. If the multiplication is changed, then the direct proportional relationship or ratio between the controlled and the uncontrolled variables is changed.

A careful examination of most boiler control applications will show that the overall control system is an interconnected matrix of the four types of control application shown in Figures 4-1 through 4-4.

Section 5
Combustion of Fuels, Excess Air, and Flue Gas Analysis

Fuels can generally be classified as gaseous, liquid, or solid. In cases where a solid fuel is finely ground, such as pulverized coal, and can be transported in an air stream, its control characteristics approach those of a gaseous fuel. Liquid fuels, as they are atomized and sprayed into a furnace, also have control characteristics similar to those of a gaseous fuel. The control treatment of a solid fuel that is not finely ground is quite different from that of a gaseous or a liquid fuel.

Whether a fuel is a gas, a liquid, or a solid is determined by the ratio of its two primary chemical ingredients, carbon and hydrogen. Natural gas has an H/C ratio in excess of 0.3. Fuel oil has an H/C ratio above 0.1 and the H/C ratio of coal is usually below 0.07. Since hydrogen is the lightest element and the molecular weight of carbon is approximately six times that of hydrogen, a decrease in the H/C ratio increases the specific gravity and the density of the fuel.

5-1 Gaseous Fuels — Their Handling and Preparation

The most used gaseous fuel is natural gas, but "waste gas" or gas produced as a process byproduct may allow the replacement of natural gas or other purchased fuels. In the iron and steel industries such gases are coke oven gas and blast furnace gas. In the petroleum refining industry, a mixture of these gases is known as refinery gas. In the petrochemical industry, such gases may be called tail gas or off gas. A characteristic of these gases is a significant difference in heat content and other physical and chemical characteristics as compared to natural gas.

Natural gases vary in their chemical analyses and thus in their heating values. While the average heating value is approximately 1000 Btu per standard cubic foot (scf), Table 5-1 shows that these gases may commonly vary from 950 to over 1100 Btu per scf. Note that in all cases over 80 percent of the gas by volume is methane.

Natural gas is the only major fuel that is delivered by the supplier as it is used. The transfer of this fuel to the user usually occurs at a metering and pressure-reducing station as shown in Figure 5-1. The pressure-reducing valves reduce the pressure from the supplier's pipeline to the pressure required in the user's boiler control system. In addition the gas is metered for billing purposes and in many cases a calorimeter is also installed for recording the heating value of the gas that is supplied. No fuel preparation is required before the gas enters the boiler plant except for the reduction in pressure described above.

In the boiler control system the regulation of fuel Btu input can be accomplished with a standard control valve. The design of the valve is based on the capacity requirement of the system; the specific gas properties, and the pressure drop available for control of the gas flow. The supply of the "waste gas" streams is usually pressure- and/or flow-controlled based on the ability of the process to produce the particular gases. In almost 100 percent of the cases, some purchased natural gas must be used to meet the total plant demand for steam.

A mixture of propane and air is a fuel alternate for natural gas. It will burn in the same burners and under the same conditions as natural gas. A typical system for mixing the propane and air is shown in Figure 5-2. There are two precautions for properly mixing and burning a propane-air mixture. Such a mixture should have a propane percentage significantly greater than 10.10 percent so that the mixture will not be explosive or flammable. Additionally, the mixture should act in the burner in much the same manner as the basic

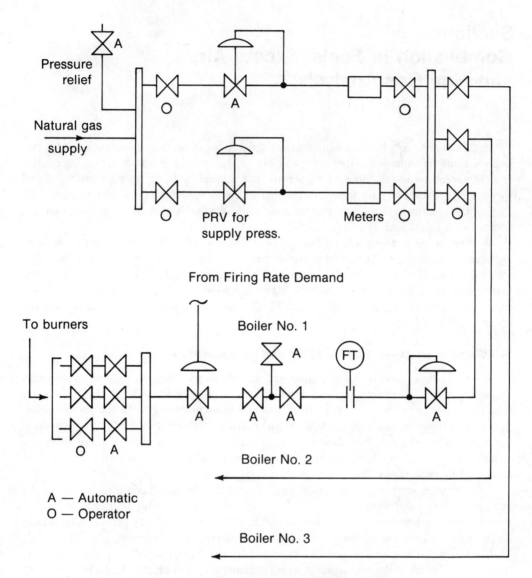

Figure 5-1 Gas Pressure — Reducing and Metering Arrangement

(From *Improving Boiler Efficiency*, Dukelow)

natural gas for which the burners are designed. Typically, the gas burner pressure for a given Btu input rate should be as close as possible to the pressure used for the basic natural gas fuel.

The characteristic properties of natural gas, air, and propane are shown in Table 5-2. For a given Btu input rate, the Btu per scf determines the number of cubic feet of gas required, and the specific gravity determines the pressure at the burner orifice.

A mixture of propane and air that will closely approximate the action of 1000 Btu per scf natural gas of 0.6 specific gravity is a mixture of 6 parts propane to 4 parts air. This produces a mixture of approximately 1500 Btu per scf with a specific gravity of 1.34. This fuel-rich mixture is 60 percent propane, well above the minimum flammable limit for a propane-air mixture.

Table 5-1
Selected Samples of Natural Gas from United States Fields

Sample No. Source of Gas		1 Pa.	2 So. Cal.	3 Ohio	4 La.	5 Okla.
Analyses						
Constituents, % by vol						
H_2	Hydrogen	—	—	1.82	—	—
Ch_4	Methane	83.40	84.00	93.33	90.00	84.10
C_2H_4	Ethylene	—	—	0.25	—	—
C_2H_6	Ethane	15.80	14.80	—	5.00	6.70
CO	Carbon monoxide	—	—	0.45	—	—
CO_2	Carbon dioxide	—	0.70	0.22	—	0.80
N_2	Nitrogen	0.80	0.50	3.40	5.00	8.40
O_2	Oxygen	—	—	0.35	—	—
H_2S	Hydrogen sulfide	—	—	0.18	—	—
Ultimate, % by wt						
S	Sulfur	—	—	0.34	—	—
H_2	Hydrogen	23.53	23.30	23.20	22.68	20.85
C	Carbon	75.25	74.72	69.12	69.26	64.84
N_2	Nitrogen	1.22	0.76	5.76	3.06	12.90
O_2	Oxygen	—	1.22	1.58	—	1.41
Specific gravity (rel to air)		0.636	0.636	0.567	0.600	0.630
Higher heat value						
Btu/cu ft @ 60° F & 30 in Hg		1,129	1,116	964	1,002	974
Btu/lb of fuel		23,170	22,904	22,077	21,824	20,160

(From *Steam, Its Generation and Use,* ©Babcock and Wilcox)

Table 5-2
Characteristic Properties

	Btu/scf	Specific Gravity
Natural gas	1000	.6
Propane	2524	1.5617
Air	0	1.0
Flammable limits — propane-to-air 2.10% to 10.10%		

5-2 Liquid Fuels — Their Handling and Preparation

The most common liquid fuel is fuel oil, a product of the oil refining process. While crude oil as produced from the well is sometimes used, the most common fuel oils used for boiler fuel are the lightweight No. 2 fuel oil and the No. 6 grade of heavy residual fuel oil. The normal range of analyses of these two fuel oils are shown in Table 5-3.

Other liquid fuels that are used as waste or auxiliary fuels are process byproducts such as tar, pitch, or acid sludge and in some cases liquid sewage. In some of these cases the heat content alone may not pay for burning the fuel and the economics may be based on a comparison with the costs of other methods of waste disposal.

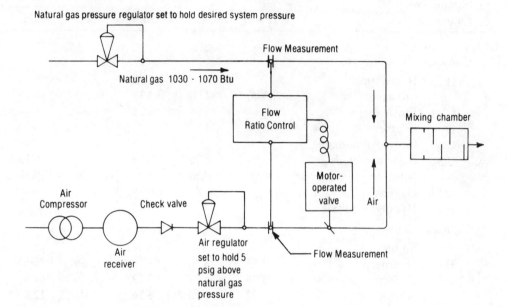

Simple stabilization control. This type of plant is suitable when the base gas flow is steady or changes very slowly — and when the heating values of constituents change very slowly.

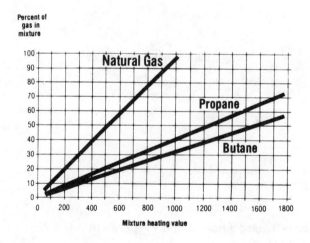

Mixture heating value versus percentage of gas in mixture for various gas-air mixtures.

Figure 5-2 A Typical System for Mixing Propane and Air

(From *Improving Boiler Efficiency*, Dukelow)

Table 5-3
Fuel Oil Analyses

	No. 2	No. 6
Carbon	86.1 to 88.2	86.5 to 90.2
Hydrogen	11.8 to 13.9	9.5 to 12.0
Sulphur	0.05 to 1.0	0.7 to 3.5
Nitrogen	Nil to 0.1	—
Ash	0	0.01 to 0.5
Heating value:		
Btu/lb	19,170 to 19,750	17,410 to 18,990
Water and sediment	0 to 0.01	0.05 to 2.0
Spec. gravity	0.887 to 0.825	1.022 to 0.922
Lb per gal	7.39 to 6.87	8.51 to 7.68

Particularly in pulp and paper mills there is a byproduct liquid fuel known as black liquor or red liquor. This liquid is burned in order to recover the chemical content of the "liquor". The heat produced from the combustion of the dissolved "wood" chemicals as the liquor is burned is a bonus. Because the basic process need is chemical recovery, the process is operated to optimize chemical recovery rather than the heat from the combustion of the wood waste content.

If the user is not an oil refinery, the fuel oil is purchased in lots and delivered to the plant by truck, railroad tank car, or oil tanker. The fuel oil is pumped from these delivery vehicles into a user's fuel oil storage tanks and stored there until used. A generic arrangement including the fuel oil preparation is shown in Figure 5-3.

In this arrangement the fuel oil is delivered to the storage tanks. From the storage tanks the fuel may be taken directly to the fuel preparation equipment, or it may be transferred to a smaller tank, sometimes called a day tank. From the day tank, fuel oil pumps provide the pressure necessary for the fuel control and atomizing system.

If the fuel is No. 2 fuel oil, heating of the fuel is normally unnecessary. If the fuel is a heavy oil such as No. 6, it is usually necessary to heat the oil in the tanks so that it can be easily pumped through the system. If heavy fuel oil in a tank is unused for a period of time, the tank heating may cause the evaporation of some of the lighter constituents, ultimately making the oil too thick to remove from the tank by any normal means.

In some installations water may be present in the oil system. This may be water that has condensed from the atmosphere over a period of time or water originating from cleaning the tanks with water. The mixture of oil and water can be burned with good results if the water is emulsified with the oil before atomizing at the burner. Emulsification forms tiny droplets of the water that are surrounded by a film of oil. As the water droplets enter the furnace, the furnace heat causes the water droplets to suddenly flash to steam — causing fine atomization of the oil film.

As a result of this type of action, water may be intentionally added to oil and emulsified to improve atomization. Due to poor atomization by existing oil burning equipment, this method has been substituted in some installations with improved results.

There is a heat loss penalty for using this method for improving atomization. Heat is lost by vaporization of the water. The heat loss is equal to the latent heat of vaporization (approximately 1040 Btu/lb of water under these conditions) multiplied by the total weight of water used.

Heavy fuel oil must also be heated before burning in order to reduce its viscosity. Figure

5-4 shows the temperature-viscosity relationship of the various grades of fuel oil. Since most burners are designed for a viscosity of 135 to 150 Saybolt universal seconds (SSU), the temperature control of the fuel oil is set to produce the desired viscosity for whatever fuel is being used. Note that specifications for No. 6 oil cover a band of viscosities. Because of this, the correct temperature set point necessary to produce the desired viscosity may vary depending upon the specific fuel characteristics.

In the boiler control system the regulation of fuel Btu input can be accomplished by a standard control valve. The design of the valve is based on the capacity requirement of the system, the specific fuel oil properties, and the pressure drop available for control of the fuel oil flow. As with the gas-fired systems, the use of the waste or auxiliary liquid fuels is usually dependent upon the ability of the process to produce such byproduct fuels. Also as in the gas-fired system, the use of such fuels may be based on disposal needs rather than the heat of combustion available.

5-3 Solid Fuels — Their Handling and Preparation

The most commonly purchased solid fuel is coal. Coal is available in a number of grades and classifications that vary from bituminous coal with almost 15,000 Btu per lb to a low-grade coal called lignite of approximately 7,000 Btu per lb. There are two different analyses of coal. One of these, called proximate analysis, is used primarily for ranking coal.

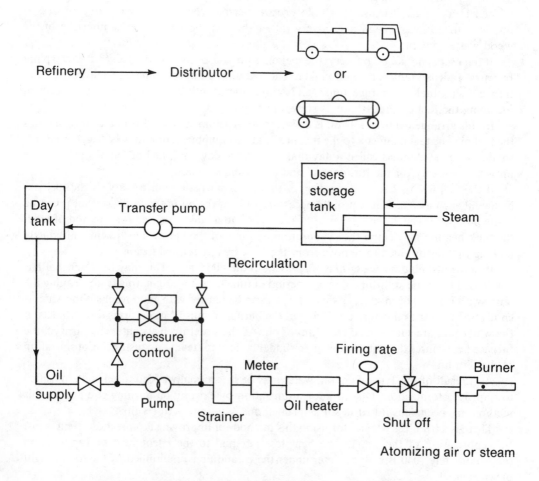

Figure 5-3 Typical Fuel Oil Pumping and Heating Arrangement

The other, called ultimate analysis, is the analysis of the chemical constituents by weight percentage. The ultimate analysis is the one used in combustion calculations. Table 5-4 demonstrates the difference between the two analysis methods. Table 5-5 shows a comparison of typical coals produced by mines in different localities.

Auxiliary or waste solid fuel is usually a process byproduct that is burned either for its heating value or for disposal purposes. The most frequently used solid auxiliary fuel is some form of wood waste from wood product manufacturing processes such as lumber saw mills and pulp and paper manufacture. Typical of these are the bark from the pulpwood that is made into paper pulp and sawdust from a sawmill. The residue of sugar cane (called bagasse) is used in the sugar industry, and coffee grounds are used in plants that manufacture instant coffee. Other solid auxiliary fuel is solid waste such as municipal garbage or other refuse.

Coal is delivered to the user in lump or chunk form by means of trucks or rail cars. The user must have sufficient space and mechanical equipment for storing and handling the coal prior to use. The fuel is often furnished on a sized basis after preliminary preparation near

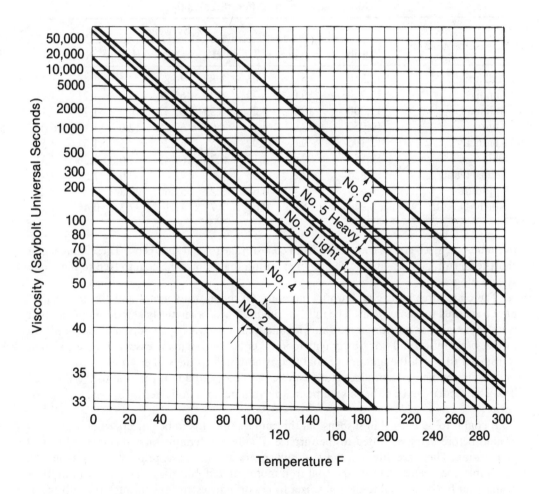

Figure 5-4 Fuel Oil Temperature vs Viscosity

(from *Steam, Its Generation and Use*, © Babcock and Wilcox)

Table 5-4
Coal Analyses On As-Received Basis
(Pittsburgh Seam Coal, West Virginia)

Proximate analysis		Ultimate analysis	
Component	Weight, %	Component	Weight, %
Moisture	2.5	Moisture	2.5
Volatile matter	37.6	Carbon	75.0
Fixed carbon	52.9	Hydrogen	5.0
Ash	7.0	Sulfur	2.3
Total	100.0	Nitrogen	1.5
		Oxygen	6.7
Heating value		Ash	7.0
Btu/lb	13,000	Total	100.0

(From *Steam, Its Generation and Use,* © Babcock and Wilcox)

Table 5-5
Coal Analysis from Different Localities

		% by weight						
	Btu lb	Ash	S	H_2	C	H_2O	N_2	O_2
Alabama	13,350	8.0	1.06	4.87	74.8	3.5	1.46	6.32
Illinois	11,200	9.0	3.79	4.26	61.78	12.0	1.21	7.96
Indiana	11,300	10.7	4.2	4.35	61.52	10.8	1.25	7.18
Kansas	10,950	12.0	4.1	4.09	60.65	12.0	1.16	6.0
Kentucky	12,100	9.0	2.92	4.5	67.24	7.5	1.32	7.52
Ohio	12,550	9.0	3.18	4.74	69.41	5.0	1.36	7.31
Pennsylvania	13,850	8.0	1.96	4.63	77.87	3.0	1.57	2.97
Colorado	9,670	5.4	0.6	3.2	57.6	20.8	1.2	11.2
North Dakota	6,940	5.9	0.6	2.7	41.1	37.3	0.4	12.0

the mine. The sized coal is transferred from coal storage to coal "bunkers". These bunkers, or short-term storage bins, are located at an elevation that allows the coal to feed to the boiler system by gravity.

There are three basic coal burning methods. Of these, "on a grate" and "in suspension" have been used for many years. The third is "fluidized bed" a relatively new development that is now emerging as a future major burning method. For grate-type burning, coal needs no further preparation and flows by gravity to a stoker hopper. In many cases a weighing system is installed to weigh the amount of coal that is admitted to the stoker hopper.

A fluidized bed system uses an adjustable rate feeder to admit the coal to the fluidized bed. Thie feeder can be either a volumetric (by volume) or a gravimetric (by weight) type of device.

If coal is to be burned in suspension, this implies that it is to be pulverized or very finely ground before being admitted to the furnace. Figure 5-5 demonstrates this method of fuel preparation. There are different types of pulverizers, but the basic principle is the same. An adjustable coal feeder delivers the coal to a motor-driven pulverizer (also called a mill). A primary air fan connected to the pulverizer forces or induces air flow through the pulverizer. As the coal is ground to a fine powder, the primary air stream transports the coal to the burner. To dry the moisture from the coal, the temperature of the coal-air mixture is controlled by mixing hot air and ambient air.

Most auxiliary or waste solid fuels are burned on a grate. The design of the grate burning system is based on the particular fuel to be burned. Some such fuels may be prepared briquettes that can be burned very similarly to coal on a coal stoker. For other fuels, depending on the condition of the fuel or the type of grate burning system used, the fuel may or may not need further preparation before being admitted to the furnace. A common preparation for wood waste is called "hogging". A hogger shreds the wood waste to a somewhat uniform size to improve the performance of the fuel burning system. After hogging the fuel is conveyed to a bin from which variable-speed screw feeders or other adjustable devices deliver the fuel to the furnace.

5.4 Fuel Mixtures — Coal-Oil, Coal-Water

To reduce fuel costs or when a preferred fuel is in short supply, fuels can be burned in combination. Burning of gaseous, liquid, or solid fuels in separate systems but in combination burning requires the same fuel preparation as if the fuel were burned alone. For such burning, boiler designs must be the same as if the fuels were burned alone.

The existence of thousands of gas- and/or oil-designed installations has spurred an effort to find ways of burning coal in boilers designed for oil firing. A mixture of finely ground coal in a colloidal mixture with oil produces a fuel that can be burned in an oil furnace in a manner very similar to that of No. 6 fuel oil. This mixture contains up to approximately 50 percent coal. Additives in the mixture help to keep the coal in suspension so that it will not

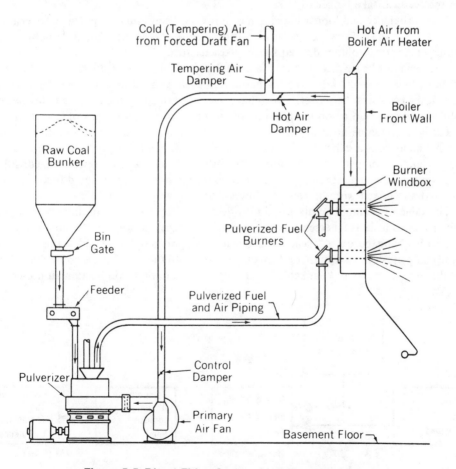

Figure 5-5 Direct Firing System for Pulverized Coal

(From *Steam, Its Generation and Use*, © Babcock and Wilcox)

settle out in the storage tanks. The advantage of this process is the reduction in the cost of fuel, since the price of coal for a given heat content is much less than that of fuel oil.

A similar developing fuel that substitutes for fuel oil is a mixture of water and pulverized coal (approximately 70 percent coal) in which the coal is kept in suspension with chemical additives. The advantage of this fuel is the complete elimination of the higher priced fuel oil. A disadvantage is the resulting lower boiler efficiency due to the increased latent heat loss of the water vapor in the flue gas. Both the coal-oil mixture and the coal-water mixture require a pulverizing system, a mixing system, and a storage and handling system similar to that for heavy fuel oil.

5-5 Physical Combustion Requirements

Combustion is the combination of fuel and air with heat produced and carried by the mass of flue gas generated. Combustion takes place, however, only under the conditions shown in Figure 5-6.

Time, temperature, and turbulence are known as the three "T"s of combustion. A short period of time, high temperature, and very turbulent flame indicate rapid combustion. Turbulence is a key because the fuel and air must be mixed thoroughly if the fuel is to be completely burned. When fuel and air are well mixed and all the fuel is burned, the flame temperature will be high and the combustion time will be shorter. When the fuel and air are not well mixed, complete combustion may not occur, the flame temperature will be lower and the fuel will take longer to burn.

Less turbulence and longer burning has, however, been found to produce fewer nitrous oxides (NOx). In some cases, combustion is delayed or staged intentionally to obtain fewer nitrous oxides or to obtain desired flame characteristics.

The fuel must be gasified. In the case of natural gas, this is automatically true. For oil, the fuel must be atomized so that the temperature present can turn it into a gas. When coal is burned, the coal must be pulverized so that it can be gasified by the furnace temperature or distilled in suspension or on the grate by the furnace temperature if a stoker is used.

The ignition temperature and also the flame temperature are different for different fuels if all other conditions are the same. Table 5-6 lists the flammable limits of the "standard" fuels and the ignition temperatures of a number of fuels. Of these, when properly mixed with air, the gases have the highest temperature required for ignition. Various liquid fuels if properly atomized and mixed with air have the lowest ignition temperatures.

The table of flammable mixture limits identifies that coal may continue burning or be ignited with as little as 8 percent of the theoretical air required for combustion. Natural gas, on the other hand, becomes "fuel-rich" and cannot be ignited or burned if less than 64 percent of the complete combustion theoretical air is mixed with the fuel. This is consistent with the fact that a coal fire can be "banked" with a very small amount of combustion air present, and

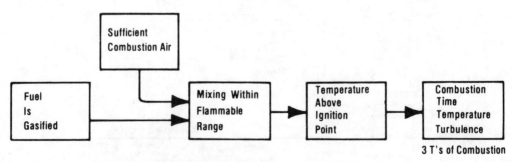

Figure 5-6 Combustion Requirements

(From *Improving Boiler Efficiency*, Dukelow)

Table 5-6
Flammable Limits

Percentage of Stoichiometric Air

	Minimum	Maximum
Natural gas	64	247
Oil	30	173
Coal (pulverized)	8	425

Ignition Temperatures of Fuels

Fuel	Degrees F
Kerosene	500
Light fuel oil	600
Gasoline	735
N-Butane	760
Heavy fuel oil	765
Coal	850
Propane	875
Natural gas	1000
Hydrogen	1095
Carbon monoxide	1170
Natural gas	1200

(From *Flame Safeguard Controls,* © Honeywell, Inc. Used with permission.)

that combustion will continue at a very low rate until a larger amount of combustion air is admitted.

Combustion may take place when the physical requirements are within the limits but may not proceed to completion due to insufficient combustion air and/or insufficient turbulence for complete mixing of the fuel and air. This can also occur if the gases are chilled by heat being withdrawn before the combustion proceeds to completion. The chilling may occur if the flame impinges on relatively cold boiler heat transfer surfaces in the furnace. It can also occur if the furnace volume is too low and allows too little time for complete combustion to take place before the gases are chilled by the convection heating surfaces of the boiler. Figure 5-7 demonstrates some of these effects as they relate to the products of combustion.

5-6 Combustion Chemistry and Products of Combustion

For all fuels, the actual chemical process is the oxidation of the hydrogen and the carbon in the fuel by combining them with oxygen from the air. The nitrogen from the air and any other non-combustibles in the fuel pass through the process with essentially no chemical change. A minimal amount of nitrogen in the air combines with oxygen to form nitrous oxides (NOx), which pollute the air. Some fuels contain a small percentage of sulphur, which — when burned — results in sulphur oxides that pollute the air and may corrode the boiler if the flue gas containing them is allowed to cool below the dew point.

Figure 5-8 demonstrates the basic chemical process and the chemical elements and compounds involved in complete and incomplete combustion. For any fuel, a precise amount of combustion air is needed to furnish the oxygen for complete combustion of that fuel's carbon and hydrogen.

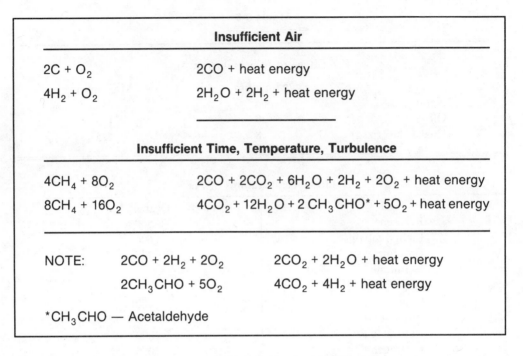

Insufficient Air

$2C + O_2$ $2CO$ + heat energy

$4H_2 + O_2$ $2H_2O + 2H_2$ + heat energy

Insufficient Time, Temperature, Turbulence

$4CH_4 + 8O_2$ $2CO + 2CO_2 + 6H_2O + 2H_2 + 2O_2$ + heat energy

$8CH_4 + 16O_2$ $4CO_2 + 12H_2O + 2\ CH_3CHO^* + 5O_2$ + heat energy

NOTE: $2CO + 2H_2 + 2O_2$ $2CO_2 + 2H_2O$ + heat energy

             $2CH_3CHO + 5O_2$ $4CO_2 + 4H_2$ + heat energy

$^*CH_3CHO$ — Acetaldehyde

Figure 5-7 Incomplete Combustion Examples

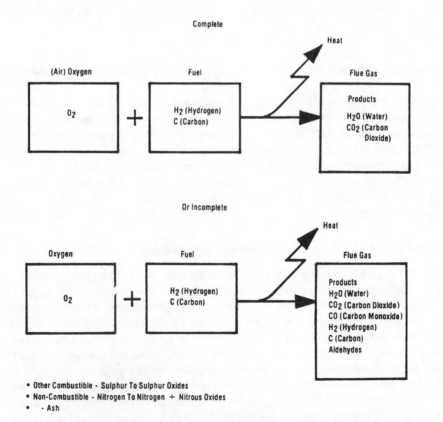

Figure 5-8 Basic Combustion Chemistry and Products of Combustion

The precise amount of combustion air is called the theoretical air for that particular fuel. If the fuel analysis is known, the theoretical air requirements can be calculated easily.

The amounts of carbon and oxygen for complete combustion of carbon are represented by the formula:

(carbon) + (oxygen)   =   (carbon dioxide)
$C + O_2$ = CO_2 + 14,100 Btu/lb C
12 lbs + 32 lbs   =   44 lbs

Weights equivalent to the molecular weight in pounds combine. One molecule of carbon containing one atom of carbon combines with one molecule of oxygen containing two atoms of oxygen to form one molecule of carbon dioxide containing one atom of carbon and two atoms of oxygen.

The formula for the combustion of hydrogen is represented:

(hydrogen) + (oxygen)   =   water
$2H_2 + O_2$ = $2H_2O$ + 61,100 Btu/lb H_2
4 lbs + 32 lbs   =   36 lbs

As with the carbon combustion, weights equivalent to the molecular weights in pounds combine. Two molecules of hydrogen, each containing two atoms of hydrogen, and one molecule of two atoms of oxygen make two molecules of water with a total of four atoms of hydrogen and two atoms of oxygen.

A simple example of the many incomplete combustion reactions resulting in intermediate hydrocarbon compounds is the partial combustion of carbon, resulting in carbon monoxide rather than carbon dioxide. In this case some of the potential heat energy from the carbon remains in the carbon monoxide.

(carbon) + (oxygen)   =   (carbon monoxide)
$2C + O_2$ =   2CO + 4,000 Btu/lb C
24 lbs + 32 lbs   =   56 lbs

With the right conditions of time, temperature, and turbulence, and by adding more oxygen to the carbon monoxide, the carbon monoxide will further oxidize to carbon dioxide, releasing the second part of the heat energy from the original carbon.

(carbon monoxide) + (oxygen)   =   (carbon dioxide)
$2CO + O_2$ =   2 CO + 4,435 Btu/lb CO
56 lbs + 32 lbs   =   88 lbs

The common chemical reactions in combustion are shown in Table 5-7, with the heat energy resulting from the combustion reaction.

Figures 5-7 and 5-8 and Table 5-7 identify those products of combustion that are produced by the oxidation of the hydrogen, carbon, or sulfur present in the fuel. As indicated, the combustion process produces heat, but a small amount of the heat produced is not useful in transferring heat to the boiler water. As hydrogen combines with oxygen during the combustion process to form water, the combustion temperature vaporizes the water into superheated steam. This vaporization absorbs the latent heat for producing the steam from the hot combustion gases. As the gases pass through the boiler and exit from the system, the gases retain the vaporized water in the form of superheated steam, and the latent heat and any remaining sensible heat is lost from the process. The amount of latent heat loss is determined by the hydrogen content of the fuel. If the fuel is natural gas and thus is higher in % hydrogen, the latent heat loss is greater than if the fuel were coal, which is lower in % hydrogen. The effect on boiler efficiency for different fuels is shown in Figure 5-9.

Since this latent heat is not useful to a combustion process, the fuel is said to have a "gross" and "net" heating value or a higher (HHV) or a lower (LHV) heating value. It is important to keep in mind that combustion air must be furnished for the total combustion or on the basis of the HHV, while only the LHV has any effect on the heat transfer of the system. Figure 5-10 demonstrates with a coal analysis how the difference between these two heating values can be calculated.

5-7 Theoretical Air Requirements and Relationship to Heat of Combustion

Using the combustion chemistry formulas, if the fuel analysis is known, the theoretical amount of oxygen can be calculated. The amount of oxygen can easily be converted to a quantity of combustion air due to the known content of oxygen in air. An example of this calculation using a formula developed from the combustion equations and the known content of oxygen in air is given in Figure 5-11. In this example the amount of air theoretically required to produce 10,000 Btu is also shown. Table 5-8 is a tabulation of combustion constants that is useful in simplifying such calculations. Figure 5-12 demonstrates using the table of combustion constants for a gaseous fuel of 85 percent methane and 15 percent ethane.

Table 5-7
Common Chemical Reactions in Combustion

Combustible	Reaction	Moles	Pounds	Heat of combustion (high) Btu/lb of fuel
Carbon (to CO)	$2C + O_2 = 2CO$	$2+1=2$	$24+ 32= 56$	4000
Carbon (to CO_2)	$C + O_2 = CO_2$	$1+1=1$	$12+ 32= 44$	14,100
Carbon monoxide	$2CO + O_2 = 2CO_2$	$2+1=2$	$56+ 32= 88$	4,345
Hydrogen	$2H_2 + O_2 = 2H_2O$	$2+1=2$	$4+ 32= 36$	61,100
Sulfur (to SO_2)	$S + O_2 = SO_2$	$1+1=1$	$32+ 32= 64$	3,980
Methane	$CH_4 + 2O_2 = CO_2 = 2H_2O$	$1+2=1+2$	$16+ 64= 80$	23,875
Acetylene	$2C_2H_2 + 5O_2 = 4CO_2 + 2H_2O$	$2+5=4+2$	$52+160=212$	21,500
Ethylene	$C_2H_4 + 3O_2 = 2CO_2 + 2H_2O$	$1+3=2+2$	$28+ 96=124$	21,635
Ethane	$2C_2H_6 + 7O_2 = 4CO_2 + 6H_2O$	$2+7=4+6$	$60+224=284$	22,325
Hydrogen sulfide	$2H_2S + 3O_2 = 2SO_2 + 2H_2O$	$2+3=2+2$	$68+ 96=164$	7,100

(From *Steam, Its Generation and Use,* © Babcock and Wilcox)

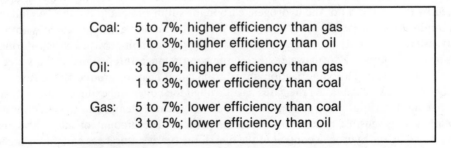

Coal:	5 to 7%; higher efficiency than gas
	1 to 3%; higher efficiency than oil
Oil:	3 to 5%; higher efficiency than gas
	1 to 3%; lower efficiency than coal
Gas:	5 to 7%; lower efficiency than coal
	3 to 5%; lower efficiency than oil

Figure 5-9 Products of Combustion and Effects on Boiler Efficiency

Table 5-8
Combustion Constants

No.	Substance	Formula	Molecular Weight	Lb per cu ft	Cu ft per lb	Sp Gr Air=1.0000	Heat of combustion Btu per cu ft Gross (High)	Net (Low)	Btu per lb Gross (High)	Net (Low)	Required for combustion (moles/cu ft) O₂	N₂	Air	Flue products CO₂	H₂O	N₂	Required for combustion (lb) O₂	N₂	Air	Flue products CO₂	H₂O	N₂
1	Carbon*	C	12.01	—	—	—	—	—	14,093	14,093	1.0	3.76	4.76	1.0	—	3.76	2.66	8.86	11.53	3.66	—	8.86
2	Hydrogen	H₂	2.016	0.0053	187.723	0.0696	325	275	61,095	51,623	0.5	1.88	2.38	—	1.0	1.88	7.94	26.41	34.34	—	8.94	26.41
3	Oxygen	O₂	32.00	0.0846	11.819	1.1053	—	—	—	—	—	—	—	—	—	—	—	—	—	—	—	—
4	Nitrogen (atm)	N₂	28.01	0.0744	13.443	0.9718	—	—	—	—	—	—	—	—	—	—	—	—	—	—	—	—
5	Carbon monoxide	CO	28.01	0.0740	13.506	0.9672	321	321	4,347	4,347	0.5	1.88	2.38	1.0	—	1.88	0.57	1.90	2.47	1.57	—	1.90
6	Carbon dioxide	CO₂	44.01	0.1170	8.548	1.5282	—	—	—	—	—	—	—	—	—	—	—	—	—	—	—	—
Paraffin series																						
7	Methane	CH₄	16.04	0.0425	23.552	0.5543	1012	911	23,875	21,495	2.0	7.53	9.53	1.0	2.0	7.53	3.99	13.28	17.27	2.74	2.25	13.28
8	Ethane	C₂H₆	30.07	0.0803	12.455	1.0488	1773	1622	22,323	20,418	3.5	13.18	16.68	2.0	3.0	13.18	3.73	12.39	16.12	2.93	1.80	12.39
9	Propane	C₃H₈	44.09	0.1196	8.365	1.5617	2524	2342	21,669	19,937	5.0	18.82	23.82	3.0	4.0	18.82	3.63	12.07	15.70	2.99	1.63	12.07
10	n-Butane	C₄H₁₀	58.12	0.1582	6.321	2.0665	3271	3018	21,321	19,678	6.5	24.47	30.97	4.0	5.0	24.47	3.58	11.91	15.49	3.03	1.55	11.91
11	Isobutane	C₄H₁₀	58.12	0.1582	6.321	2.0665	3261	3009	21,271	19,628	6.5	24.47	30.97	4.0	5.0	24.47	3.58	11.91	15.49	3.03	1.55	11.91
12	n-Pentane	C₅H₁₂	72.15	0.1904	5.252	2.4872	4020	3717	21,095	19,507	8.0	30.11	38.11	5.0	6.0	30.11	3.55	11.81	15.35	3.05	1.50	11.81
13	Isopentane	C₅H₁₂	72.15	0.1904	5.252	2.4872	4011	3708	21,047	19,459	8.0	30.11	38.11	5.0	6.0	30.11	3.55	11.81	15.35	3.05	1.50	11.81
14	Neopentane	C₅H₁₂	72.15	0.1904	5.252	2.4872	3994	3692	20,978	19,390	8.0	30.11	38.11	5.0	6.0	30.11	3.55	11.81	15.35	3.05	1.50	11.81
15	n-Hexane	C₆H₁₄	86.17	0.2274	4.398	2.9704	4768	4415	20,966	19,415	9.5	35.76	45.26	6.0	7.0	35.76	3.53	11.74	15.27	3.06	1.46	11.74
Olefin series																						
16	Ethylene	C₂H₄	28.05	0.0742	13.475	0.9740	1604	1503	21,636	20,275	3.0	11.29	14.29	2.0	2.0	11.29	3.42	11.39	14.81	3.14	1.29	11.39
17	Propylene	C₃H₆	42.08	0.1110	9.007	1.4504	2340	2188	21,048	19,687	4.5	16.94	21.44	3.0	3.0	16.94	3.42	11.39	14.81	3.14	1.29	11.39
18	N-Butene	C₄H₈	56.10	0.1480	6.756	1.9336	3084	2885	20,854	19,493	6.0	22.59	28.59	4.0	4.0	22.59	3.42	11.39	14.81	3.14	1.29	11.39
19	Isobutene	C₄H₈	56.10	0.1480	6.756	1.9336	3069	2868	20,737	19,376	6.0	22.59	28.59	4.0	4.0	22.59	3.42	11.39	14.81	3.14	1.29	11.39
20	n-Pentene	C₅H₁₀	70.13	0.1852	5.400	2.4190	3837	3585	20,720	19,359	7.5	28.23	35.73	5.0	5.0	28.23	3.42	11.39	14.81	3.14	1.29	11.39
Aromatic series																						
21	Benzene	C₆H₆	78.11	0.2060	4.852	2.6920	3752	3601	18,184	17,451	7.5	28.23	35.73	6.0	3.0	28.23	3.07	10.22	13.30	3.38	0.69	10.22
22	Toluene	C₇H₈	92.13	0.2431	4.113	3.1760	4486	4285	18,501	17,672	9.0	33.88	42.88	7.0	4.0	33.88	3.13	10.40	13.53	3.34	0.78	10.40
23	Xylene	C₈H₁₀	106.16	0.2803	3.567	3.6618	5230	4980	18,650	17,760	10.5	39.52	50.02	8.0	5.0	39.52	3.17	10.53	13.70	3.32	0.85	10.53
Miscellaneous gases																						
24	Acetylene	C₂H₂	26.04	0.0697	14.344	0.9107	1477	1426	21,502	20,769	2.5	9.41	11.91	2.0	1.0	9.41	3.07	10.22	13.30	3.38	0.69	10.22
25	Napthalene	C₁₀H₈	128.16	0.3384	2.955	4.4208	5854	5654	17,303	16,708	12.0	45.17	57.17	10.0	4.0	45.17	3.00	9.97	12.96	3.43	0.56	9.97
26	Methyl alcohol	CH₃OH	32.04	0.0846	11.820	1.1052	868	767	10,258	9,066	1.5	5.65	7.15	1.0	2.0	5.65	1.50	4.98	6.48	1.37	1.13	4.98
27	Ethyl alcohol	C₂H₅OH	46.07	0.1216	8.221	1.5890	1600	1449	13,161	11,917	3.0	11.29	14.29	2.0	3.0	11.29	2.08	6.93	9.02	1.92	1.17	6.93
28	Ammonia	NH₃	17.03	0.0456	21.914	0.5961	441	364	9,667	7,985	0.75	2.82	3.57	—	1.5	3.32	1.41	4.69	6.10	—	1.59	5.51
29	Sulfur*	S	32.06	—	—	—	—	—	3,980	3,980	1.0	3.76	4.76	SO₂ 1.0	—	3.76	1.00	3.29	4.29	SO₂ 2.00	—	3.29
30	Hydrogen sulfide	H₂S	34.08	0.0911	10.979	1.1898	646	595	7,097	6,537	1.5	5.65	7.15	1.0	1.0	5.65	1.41	4.69	6.10	1.88	0.53	4.69
31	Sulfur dioxide	SO₂	64.06	0.1733	5.770	2.2640	—	—	—	—	—	—	—	—	—	—	—	—	—	—	—	—
32	Water vapor	H₂O	18.02	0.0476	21.017	0.6215	—	—	—	—	—	—	—	—	—	—	—	—	—	—	—	—
33	Air	—	—	0.0766	13.063	1.0000	—	—	—	—	—	—	—	—	—	—	—	—	—	—	—	—

*Carbon and sulfur are considered as gases for molal calculations only.

All gas volumes corrected to 60° F and 30 in. Hg dry.

(From *Steam, Its Generation and Use*, © Babcock and Wilcox)

- Gross Btu/lb as measured by calorimeter, 14,100 Btu/lb

  Ultimate analysis;   C   —   80.31%
  $\qquad\qquad\qquad\quad$ H_2 —   4.47%
  $\qquad\qquad\qquad\quad$ S   —   2.85%
  $\qquad\qquad\qquad\quad$ N_2 —   1.38%
  $\qquad\qquad\qquad\quad$ H_2O —   2.9%
  $\qquad\qquad\qquad\quad$ Ash   —   6.55%

- BUT 0.0447 lb H_2 forms water vapor and 0.029 lb water is vaporized. All water vapor discharged in flue gas.

- Water vapor formed from H_2: $0.0447 \times \dfrac{18.02}{2.016}$ = 0.3996 lb

  $\qquad\qquad\qquad\qquad\quad$ H_2O in fuel   . . . .   = 0.029 lb
  $\rule{3cm}{0.4pt}$

  $\qquad\qquad\qquad\quad$ Total H_2O/lb in fuel   . . . .   = 0.029 lb

- Heat loss due to vaporizing H_2O: $0.4286 \times 1040^*$ = 445.7 Btu

  $\qquad\qquad$ Net Btu/lb = Gross Btu/lb − Loss vaporizing H_2O or
  $\qquad\qquad\qquad\qquad$ 14,100 − 445.7 = 13.654.3 Btu/lb

- Loss vaporizing H_2O = 445.7/14,100 = 3.16%

*Standard ASTM procedure uses 1040. Actually, value changes with fuel analysis due to change in partial pressure of H_2O in the flue gases.

Figure 5-10 Gross and Net Btu (higher and lower heating value)

Note that the amount of combustion air required to produce 10,000 Btu is nearly the same for coal and natural gas. If the reciprocals are taken, the result is Btu/lb of air. Table 5-9 shows that for coal, oil, or gas the Btu/lb of air is approximately the same even though the Btu/lb of the fuels is completely different. The difference between the Btu/lb of air on a "net" basis for these fuels is smaller than that shown in the table. The fact that combustion air requirements can be closely approximate, based on the heat requirement, is an important concept used in the application of combustion control logic.

Table 5-9
Fuel and Air Heating Value Comparison

Fuel	Btu/lb Fuel	Btu/lb Air
Bituminous Coal	12,975	1,332
Subbituminous Coal	9,901	1,323
#6 Oil	18,560	1,351
#2 Oil	19,410	1,376
Natural Gas	23,170	1,393

(A) Theoretical Requirements for Combustion Air

- Lbs air/lb fuel $= 11.53\, C + 34.34 \left(H_2 - \dfrac{O_2}{8} \right) + 4.29\, S$

for 10,000 Btu

Lb air/10,000 Btu $= \dfrac{10,000 \left[11.53C + 34.34 \left(H_2 - \dfrac{O_2}{8} \right) + 4.29\, S \right]}{Btu/lb}$

C, H_2, O_2, S are decimal equivalent to ultimate analysis % by weight.

(B) Theoretical Air Example

- Assume coal; Ultimate analysis:

C	—	80.31%
H$_2$	—	4.47%
S	—	2.85%
N$_2$	—	1.38%
H$_2$O	—	2.9%
Ash	—	6.55%

- Lbs air/lb fuel $=$

$(11.53 \times 0.8031) + 34.34 \left(0.0445 - \dfrac{0.0285}{8} \right) + (4.29 \times 0.0154)$

$9.259 + 1.413 + 0.0661 = 10.7381$

- Lbs air/10,000 Btu $= \dfrac{10,000 \times 10.7381}{14,100} = 7.6156$

Figure 5-11 Combustion Equations and Theoretical Air Examples

5-8 The Requirement of Excess Combustion Air

In actual practice gas-, oil-, coal-burning, and other systems do not do a perfect job of mixing the fuel and air even under the best achievable conditions of turbulence. Additionally, complete mixing may take too much time — so that the gases pass to a lower temperature area not hot enough to complete the combustion — before the process is completed.

If only the amount of theoretical air were furnished, some fuel would not burn, the combustion would be incomplete, and the heat in the unburned fuel would be lost. To assure complete combustion, additional combustion air is furnished so that every molecule of the fuel can easily find the proper number of oxygen molecules to complete the combustion.

This additional amount of combustion air furnished to complete the combustion process is called excess air. Excess air plus theoretical air is called total air. Having this necessary excess air means that some of the oxygen will not be used and will leave the boiler in the flue gases as shown in Figure 3-4 which describes the fuel and air chemicals' mass balance. The oxygen portion of the flue gas can be used to determine the percentage of excess air.

Theeoretical Air Example

Fuel – natural gas: 85% methane, 15% ethane

Btu/scf = (0.85 × 1012) + (0.15 × 1773) = 1126.15

Cu ft air required/cu ft fuel   = (0.85 × 9.53) + (0.15 × 16.68) = 10.607

Cu ft air required/10000 Btu = 10.607 × (10000/1126.15) = 94.277

Lbs air required/10000 Btu   = 94.277/13.063 = 7.217

Btu/lb air = 0000/7.217 = 1386 Btu

**Figure 5-12 Theoretical Air Calculation by Using
Table of Combustion Constants**

If the percentage of excess air is increased, flame temperature is reduced and the boiler heat transfer rate is reduced. The usual effect of this change is the increase in the flue gas temperature as shown in Figure 5-13.

If oxygen in the flue gas is known or can be measured and no table or curve is available, the following empirical formula provides a close approximation of the percentage of excess air. This formula is based on "dry basis" percentage oxygen.

$$\text{Excess air } (\%) \quad = \quad K \left(\frac{21}{21 - \% \text{ oxygen}} - 1 \right) \times 100$$

where K =   0.9 for gas
0.94 for oil
0.97 for coal

Measurements of either percentage of carbon dioxide or the percentage of oxygen in the flue gas or both are used to determine percentage of excess air, but the percentage of oxygen is preferred for the following reasons:

(1) Oxygen is part of the air — if oxygen is zero then excess air is zero. The presence of oxygen always indicates that some percentage of excess air is present.

(2) The percent of carbon dioxide rises to a maximum at minimum excess air and then decreases as air is further reduced. It is thus possible, with the same percentage of carbon dioxide, to have two different percentages of total combustion air. For this reason, the percentage of carbon dioxide cannot be used alone as a flue gas analysis input to a combustion control system.

(3) To determine excess air with the same precision, greater precision of measurement is required for the percentage of carbon dioxide method than for the percentage of oxygen method.

(4) The relationship between the percentage of oxygen and the percentage of excess air changes little as fuel analysis or type of fuel changes, while the percentage of carbon dioxide-to-excess air relationship varies considerably as the percentage of carbon-to-hydrogen ratio of the fuel changes.

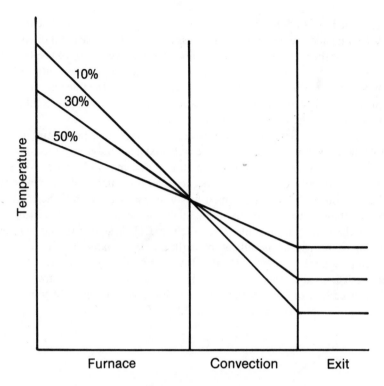

- Reduced excess air improves heat transfer.
- Reduced excess air reduces mass of flue gases.

Figure 5-13 Effect of Excess Air on Temperatures

The heat loss in the flue gases essentially depends upon the temperature of the flue gases and the incoming combustion air, the amount of the excess air, and the fuel analysis. There is an optimum amount of excess air because less air will mean unburned fuel from incomplete combustion, and more air will mean complete combustion but more heat loss in the flue gas due to the greater mass of the flue gases.

The amount of excess air required depends upon the type of fuel, burner design, fuel characteristics and preparation, furnace design, capacity as a percent of maximum, and other factors. The amount for any installation should be determined by testing that particular unit. An approximate amount of excess air required for full capacity is shown in Table 5-10.

Table 5-10
Excess Air Required at Full Capacity

Fuel	% Oxygen in flue gas	% excess air min.
Natural gas	1.5 to 3	7–15
Fuel oil	0.6 to 3	3–15
Coal	4.0 to 6.5	25–40

The charts in Figures 5-14, 5-15, and 5-16 show the relationship between the flue gas analysis by volume and the percentage of excess air for natural gas, fuel oil, and coal. While these curves are for fuels with specific fuel analyses, the curves for % oxygen vs excess air are quite similar, while the % carbon dioxide vs excess air curves are quite different for the different fuels. These curves also show the difference between the flue gas analysis depending upon the presence or removal of the water vapor that is formed by the combustion process. This difference is important to note for two reasons.

Since approximately 1970, flue gas analyzers for % oxygen using the zirconium oxide fuel cell principle have been marketed. This type of % oxygen analyzer, which analyzes the flue gas on the "wet" basis, is now the standard method for permanently installed flue gas analysis equipment. On the other hand, % oxygen vs excess air formulas, including the one in the text above, are based on the "dry" basis which was universally used until the early 1970s.

In addition, these newer zirconium oxide analyzers normally measure the % oxygen on a "net" basis. If there are combustible gases such as CO present, the high temperature and catalytic action of the measuring cell complete the combustion by subtracting a portion or all of the oxygen passing the analysis cell. It is thus not necessary to subtract CO_2 from the % oxygen before it is used in the older formulas. Since these formulas are on the dry basis, however, it is necessary to convert the wet basis analyzer readings to dry basis before using them in the older equations for combustion calculations.

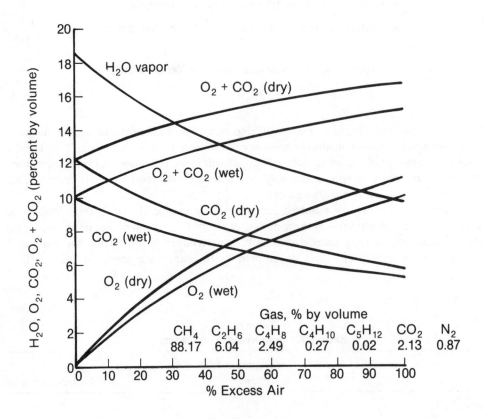

Figure 5-14. Flue Gas Composition as a Function of Excess Air for Natural Gas Fuel (Heating value approximately to 22,810 Btu/lb)

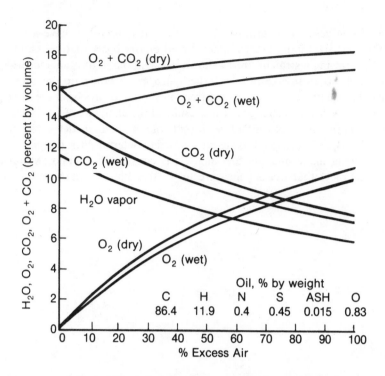

**Figure 5-15. Flue Gas Composition as a Function of Excess Air for Oil Fuels
(Heating value is approximately 18,700 Btu/lb)**

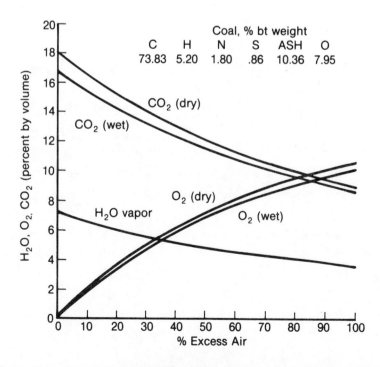

**Figure 5-16. Flue Gas Composition as a Function of Excess Air for Coal Fuels
(Heating value is approximately 13,320 Btu/lb)**

In analyzing flue gases to determine % excess air, it is useful to have a check on the accuracy of the analysis. Figure 5-17, which is based on dry basis analysis, can be used for this purpose. By drawing a straight line as shown between the % oxygen and the % carbon dioxide values, there is an intersection with the hydrogen-carbon ratio line. When using a particular fuel of a certain H/C ratio, the intersection should always be at the same point on the line. If it is not, the fuel has changed or the results of the analysis are incorrect. If the fuel analysis is known, the measurement of % oxygen can be used to determine the correct % carbon dioxide, or the % carbon dioxide can be used to determine the correct % oxygen. If boiler tests are being made and the fuel analysis is constant, any data (within limits of measurement accuracy) that doesn't measure up to this kind of examination should be thrown out.

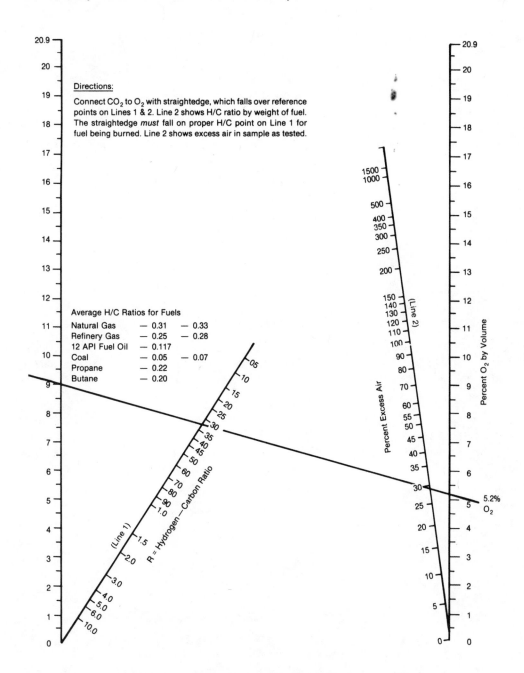

Directions:

Connect CO_2 to O_2 with straightedge, which falls over reference points on Lines 1 & 2. Line 2 shows H/C ratio by weight of fuel. The straightedge *must* fall on proper H/C point on Line 1 for fuel being burned. Line 2 shows excess air in sample as tested.

Average H/C Ratios for Fuels

Natural Gas	— 0.31	— 0.33
Refinery Gas	— 0.25	— 0.28
12 API Fuel Oil	— 0.117	
Coal	— 0.05	— 0.07
Propane	— 0.22	
Butane	— 0.20	

Figure 5-17. Properties of the Products of Combustion

Section 6
Efficiency Calculations

Two different methods for calculating the efficiency of boilers are used. Both of these methods are useful, but it is generally agreed that the heat loss method provides more accurate results. The input/output or direct method is more easily understood and for that reason is often preferred.

6-1 Input/Output or Direct Method

The basic calculation for the input/output method is shown in Figure 6-1. This formula is based on a knowledge of the flows of steam from the boiler, feedwater to the boiler, the blowdown from the boiler, and the flow of fuel to the boiler. In addition the unit heating value of each of these flows must be known.

All the flows above can be determined from flowmeters. If steam flow alone or feedwater flow alone is known, blowdown flow can be determined from the chemical relationships between feedwater entering the boiler and the water in the boiler.

If the pressure, temperature or % moisture of the steam is known, the heat content of the steam, feedwater, or blowdown can be determined from the "Thermodynamic Properties of Steam and Water" tables or a suitable computer program. The heat content of the fuel must be determined by the use of a fuel calorimeter. For gaseous fuels the calorimeter may be either a laboratory device or an on-line recording and/or indicating instrument. For liquid or solid fuels such as oil or coal, a laboratory test is necessary.

In the formula shown in Figure 6-1, the output of the boiler is credited with all the heat that is added to the incoming feedwater. This heat added includes the non-useful heat added to the blowdown flow in increasing its temperature from feedwater temperature to the saturation temperature of the boiler water.

- EFFICIENCY $= \dfrac{\text{Heat Added to Incoming Feedwater}}{\text{Heat Input (Fuel)}}$

Heat Added to Steam-Water circuit:

(Steam Flow) (Steam enthalpy–Feedwater enthalpy) +
(Blowdown Flow) (Blowdown enthalpy–Feedwater enthalpy)

Heat Input:

(Fuel Flow) (Fuel Higher Heating Value)

- LIMITATIONS

Accuracy of flow meters

Accuracy of fuel weight (solid fuel)

Heat content of fuel (measurement accuracy)

Figure 6-1 Input-Output

A minimum of three flows are used in this calculation method. The best normal operation flowmeter accuracy that can be expected for an individual meter is plus or minus 1–2%. The potential for erroneous results because of flowmeter inaccuracy is the primary weakness of this calculation method. In addition to flowmeter inaccuracy, more inaccuracies can be introduced in the determination of the heat content of the fuel.

It is also normal that a steam, water, or fuel flowmeter may not be operating under its design condition of pressure, temperature, specific gravity, etc. Unless particular care is taken in correcting the flowmeter results to design conditions, further inaccuracies can easily be introduced. With gas flowmeters, the basic design conditions of the meter may not match the conditions of the calorimeter. Particular care must be taken to see that the heating value and fuel flow are on the same basis.

Figure 6-2 is a worksheet for determination of boiler efficiency by the input/output method. Note that completion of this worksheet forces correction to design conditions for the flowmeters and calorimetric determination. The three formulas shown are based on different combinations of flowmeter results and the particular flowmeters that are available.

The usefulness of this calculation method is derived from its ease of understanding and its ease of calculation. Though precise results may not be attainable, inaccuracies will probably repeat closely from day to day. This method is therefore useful in tracking daily performance.

A.   Steam Flow, lbs   _____   Design_____
B.   Pressure, psig   _____
C.   Temperature, °F   _____
D.   Corrected Steam Flow, lbs   _____
E.   Enthalpy of Steam, Btu/lb   _____
F.   Feedwater Flow, lbs   _____   Design_____
G.   Feedwater Temperature, °F   _____
H.   Corrected Feedwater Flow, lbs   _____
I.   Enthalpy of Feedwater, Btu/lb   _____
J.   Blowdown Flow, lbs*   _____   Design_____
K.   Blowdown Temperature, °F   _____
L.   Corrected Blowdown Flow, lbs   _____
M.   Enthalpy of Blowdown, Btu/lb   _____
N.   Gas Flow, scf at std.   _____   Design_____
O.   Gas Pressure, psig   _____
P.   Gas Temperature, °F   _____
Q.   Corrected Gas Flow, scf at std. _____
R.   Heat Content of Gas at std.   _____

$$\% \text{ Efficiency} = \frac{(D)(E) + (L)(M) - (H)(I)}{(Q)(R)} \times 100 \text{ or}$$

$$\% \text{ Efficiency} = \frac{(D)(E-I) + (L)(M-I)}{(Q)(R)} \times 100 \text{ or}$$

$$\% \text{ Efficiency} = \frac{(H-L)(E-I) + (L)(M-I)}{(Q)(R)}$$

* Blowdown is measured, (H)–(D), or computed from chemical analysis.

Figure 6-2 Boiler Efficiency, Input-Output

6-2 Heat Loss Method

The heat loss method is used to calculate individual losses, totalize them, and determine boiler efficiency by subtracting the total losses from 100%. The individual losses are identified in Figure 6-3.

Figure 6-3 identifies the method of determining the losses. In this method, a particular fuel unit, 1 mole or 100 moles, 1 lb or 100 lbs of fuel, is used. For the fuel unit the mass of the flue gases is determined. The sensible heat losses are determined by multiplying the flue gas mass flows by the mean specific heat (Mcp) between the flue gas temperature and the combustion air temperature. The result is then multiplied by the difference in temperature between the flue gases and the combustion air.

Latent heat losses are determined by multiplying the mass flow of the water that is vaporized (fuel moisture plus moisture from the combustion of hydrogen) by an average figure of 1040 Btu/lb. Each loss is calculated in terms of heat units (Btu, for example), the heat units totalized, and the percentage of the heating value of the total fuel unit determined.

For calculation of the heat losses involved, there are two accepted methods. These are the "mole" method, and the "weight" method. Because of the greater involvement of the combustion chemistry in the "mole" method, many engineers who are not trained in chemistry prefer the "weight" method. Almost identical results can be obtained from either method if the calculations are carefully made.

The "weight" method is used in the "Indirect Method" of the ASME Power Test Code. The ASME Power Test Code is generally used for contractual purposes. Its use does not require the knowledge of combustion chemistry and can be accomplished with a "cookbook" approach using the standardized forms that are shown in Figures 6-4 and 6-5. Note that in the formula in line no. 65 a standard mean specific heat of 0.24 for the dry flue gas is used.

Calculate Various Losses and Subtract from 100%

- Sensible heat loss in dry flue gas:
  Mass (dry gas) $\times \Delta t \times M_{cp}$*
- Sensible heat loss from H_2O in combustion air:
  Mass (air moisture) $\times M_{cp} \times \Delta t$
- Sensible heat loss from H_2O in fuel:
  Mass (H_2O in fuel) $\times M_{cp} \times \Delta t$
- Latent heat loss from H_2O in the fuel:
  Mass (H_2O in fuel) $\times$ 1040**
- Latent heat loss from H_2O formed by combustion of H_2:
  Mass (water vapor) $\times$ 1040
- Heat loss from unburned carbon in the refuse:
  Mass (per fuel unit) $\times$ 14,100 Btu/lb
- Heat loss from unburned combustible gas in flue gases:
  Mass (per fuel unit) $\times$ Btu content specific gases

  Plus

- Radiation and unaccounted for losses

  * M_{cp} = mean specific heat
  ** Standard value of Btu/lb

Figure 6-3 Heat Loss Method

ASME TEST FORM
FOR ABBREVIATED EFFICIENCY TEST

SUMMARY SHEET   PTC 4.1-a (1964)

	TEST NO.	BOILER NO.	DATE
OWNER OF PLANT	LOCATION		
TEST CONDUCTED BY	OBJECTIVE OF TEST		DURATION
BOILER, MAKE & TYPE	RATED CAPACITY		
STOKER, TYPE & SIZE			
PULVERIZER, TYPE & SIZE	BURNER, TYPE & SIZE		
FUEL USED MINE	COUNTY	STATE	SIZE AS FIRED

	PRESSURES & TEMPERATURES				FUEL DATA				
1	STEAM PRESSURE IN BOILER DRUM	psia		COAL AS FIRED PROX. ANALYSIS	% wt		OIL		
2	STEAM PRESSURE AT S. H. OUTLET	psia		37	MOISTURE		51	FLASH POINT F*	
3	STEAM PRESSURE AT R. H. INLET	psia		38	VOL MATTER		52	Sp. Gravity Deg. API*	
4	STEAM PRESSURE AT R. H. OUTLET	psia		39	FIXED CARBON		53	VISCOSITY AT SSU* BURNER SSF	
5	STEAM TEMPERATURE AT S. H. OUTLET	F		40	ASH		44	TOTAL HYDROGEN % wt	
6	STEAM TEMPERATURE AT R.H. INLET	F			TOTAL		41	Btu per lb	
7	STEAM TEMPERATURE AT R.H. OUTLET	F		41	Btu per lb AS FIRED				
8	WATER TEMP. ENTERING (ECON.)(BOILER)	F		42	ASH SOFT TEMP.* ASTM METHOD			GAS	% VOL
9	STEAM QUALITY % MOISTURE OR P.P.M.			COAL OR OIL AS FIRED ULTIMATE ANALYSIS			54	CO	
10	AIR TEMP. AROUND BOILER (AMBIENT)	F		43	CARBON		55	CH₄ METHANE	
11	TEMP. AIR FOR COMBUSTION (This is Reference Temperature) †	F		44	HYDROGEN		56	C₂H₂ ACETYLENE	
12	TEMPERATURE OF FUEL	F		45	OXYGEN		57	C₂H₄ ETHYLENE	
13	GAS TEMP. LEAVING (Boiler) (Econ.) (Air Htr.)	F		46	NITROGEN		58	C₂H₆ ETHANE	
14	GAS TEMP. ENTERING AH (If conditions to be corrected to guarantee)	F		47	SULPHUR		59	H₂S	
	UNIT QUANTITIES			40	ASH		60	CO₂	
15	ENTHALPY OF SAT. LIQUID (TOTAL HEAT)	Btu/lb		37	MOISTURE		61	H₂ HYDROGEN	
16	ENTHALPY OF (SATURATED) (SUPERHEATED) STM.	Btu/lb			TOTAL			TOTAL	
17	ENTHALPY OF SAT. FEED TO (BOILER) (ECON.)	Btu/lb		COAL PULVERIZATION			TOTAL HYDROGEN % wt		
18	ENTHALPY OF REHEATED STEAM R. H. INLET	Btu/lb		48	GRINDABILITY INDEX*		62	DENSITY 68 F ATM. PRESS.	
19	ENTHALPY OF REHEATED STEAM R. H. OUTLET	Btu/lb		49	FINENESS % THRU 50 M*		63	Btu PER CU FT	
20	HEAT ABS/LB OF STEAM (ITEM 16 – ITEM 17)	Btu/lb		50	FINENESS % THRU 200 M*		41	Btu PER LB	
21	HEAT ABS/LB R.H. STEAM (ITEM 19 – ITEM 18)	Btu/lb		64	INPUT-OUTPUT EFFICIENCY OF UNIT %		ITEM 31 × 100 ITEM 29		
22	DRY REFUSE (ASH PIT + FLY ASH) PER LB AS FIRED FUEL	lb/lb			HEAT LOSS EFFICIENCY		Btu/lb A. F. FUEL	% of A. F. FUEL	
23	Btu PER LB IN REFUSE (WEIGHTED AVERAGE)	Btu/lb		65	HEAT LOSS DUE TO DRY GAS				
24	CARBON BURNED PER LB AS FIRED FUEL	lb/lb		66	HEAT LOSS DUE TO MOISTURE IN FUEL				
25	DRY GAS PER LB AS FIRED FUEL BURNED	lb/lb		67	HEAT LOSS DUE TO H₂O FROM COMB. OF H₂				
	HOURLY QUANTITIES			68	HEAT LOSS DUE TO COMBUST. IN REFUSE				
26	ACTUAL WATER EVAPORATED	lb/hr		69	HEAT LOSS DUE TO RADIATION				
27	REHEAT STEAM FLOW	lb/hr		70	UNMEASURED LOSSES				
28	RATE OF FUEL FIRING (AS FIRED wt)	lb/hr		71	TOTAL				
29	TOTAL HEAT INPUT (Item 28 × Item 41)/1000	kB/hr		72	EFFICIENCY = (100 – Item 71)				
30	HEAT OUTPUT IN BLOW-DOWN WATER	kB/hr							
31	TOTAL HEAT OUTPUT (Item 26×Item 20)+(Item 27×Item 21)+Item 30 /1000	kB/hr							
	FLUE GAS ANAL. (BOILER)(ECON) (AIR HTR) OUTLET								
32	CO₂	% VOL							
33	O₂	% VOL							
34	CO	% VOL							
35	N₂ (BY DIFFERENCE)	% VOL							
36	EXCESS AIR	%							

Figure 6-4 Summary Sheet, ASME Test Form for Abbreviated Efficiency Test

With the "mole" method, the combustion chemistry formulas are used to determine the number of moles of each flue gas constituent, and the individual specific heat of that constituent for the particular flue gas and combustion air temperature difference is also used. For this reason, the "mole" method is slightly more precise than the standard ASME method shown. If a precisely calculated specific heat value is used instead of the standard 0.24 value, then the overall result of the computed losses is of similar precision. For precise results both

CALCULATION SHEET — ASME TEST FORM FOR ABBREVIATED EFFICIENCY TEST — PTC 4.1-b (1964)

OWNER OF PLANT | TEST NO. | BOILER NO. | DATE

Item		
30	HEAT OUTPUT IN BOILER BLOW-DOWN WATER = LB OF WATER BLOW-DOWN PER HR x [(ITEM 15 − ITEM 17) / 1000] = ...	kB/hr
24	*If impractical to weigh refuse, this item can be estimated as follows* DRY REFUSE PER LB OF AS FIRED FUEL = % ASH IN AS FIRED COAL / (100 − % COMB. IN REFUSE SAMPLE) CARBON BURNED PER LB AS FIRED FUEL = ITEM 43 / 100 − [(ITEM 22 x ITEM 23) / 14,500] = ...	NOTE: IF FLUE DUST & ASH PIT REFUSE DIFFER MATERIALLY IN COMBUSTIBLE CONTENT, THEY SHOULD BE ESTIMATED SEPARATELY. SEE SECTION 7, COMPUTATIONS.
25	DRY GAS PER LB AS FIRED FUEL BURNED = $\frac{11CO_2 + 8O_2 + 7(N_2 + CO)}{3(CO_2 + CO)}$ x (LB CARBON BURNED PER LB AS FIRED FUEL + $\frac{3}{8}$ S) = $\frac{11 \times \text{ITEM 32} + 8 \times \text{ITEM 33} + 7(\text{ITEM 35} + \text{ITEM 34})}{3 \times (\text{ITEM 32} + \text{ITEM 34})}$ x [ITEM 24 + ITEM 47 / 2.67] = ...	
36	EXCESS AIR† = 100 x $\frac{O_2 - \frac{CO}{2}}{.2682N_2 - (O_2 - \frac{CO}{2})}$ = 100 x $\frac{\text{ITEM 33} - \frac{\text{ITEM 34}}{2}}{.2682\ (\text{ITEM 35}) - (\text{ITEM 33} - \frac{\text{ITEM 34}}{2})}$ = ...	

	HEAT LOSS EFFICIENCY	Btu/lb AS FIRED FUEL	LOSS/HHV x 100 =	LOSS %
65	HEAT LOSS DUE TO DRY GAS = LB DRY GAS PER LB AS FIRED FUEL x C_p x ($t_{lvg} - t_{air}$) = ITEM 25 x 0.24 x (ITEM 13) − (ITEM 11) = ... Unit	......	65/41 x 100 =	......
66	HEAT LOSS DUE TO MOISTURE IN FUEL = LB H_2O PER LB AS FIRED FUEL x [(ENTHALPY OF VAPOR AT 1 PSIA & T GAS LVG) − (ENTHALPY OF LIQUID AT T AIR)] = ITEM 37 / 100 x [(ENTHALPY OF VAPOR AT 1 PSIA & T ITEM 13) − (ENTHALPY OF LIQUID AT T ITEM 11)] = ...	......	66/41 x 100 =	......
67	HEAT LOSS DUE TO H_2O FROM COMB. OF H_2 = $9H_2$ x [(ENTHALPY OF VAPOR AT 1 PSIA & T GAS LVG) − (ENTHALPY OF LIQUID AT T AIR)] = 9 x ITEM 44 / 100 x [(ENTHALPY OF VAPOR AT 1 PSIA & T ITEM 13) − (ENTHALPY OF LIQUID AT T ITEM 11)] = ...	......	67/41 x 100 =	......
68	HEAT LOSS DUE TO COMBUSTIBLE IN REFUSE = ITEM 22 x ITEM 23 = ...	......	68/41 x 100 =	......
69	HEAT LOSS DUE TO RADIATION* = TOTAL BTU RADIATION LOSS PER HR / (LB AS FIRED FUEL − ITEM 28)	......	69/41 x 100 =	......
70	UNMEASURED LOSSES **	......	70/41 x 100 =	......
71	TOTAL	......	......	......
72	EFFICIENCY = (100 − ITEM 71)	......	......	......

Figure 6-5 Calculation Sheet, ASME Test Form for Abbreviated Efficiency Test

Start/Oil Firing

Component	Lbs/ Fuel Unit	Mol Wt. Div.	Moles Fuel Unit	O_2 Mult.	O_2 Moles	CO_2 SO_2	O_2	N_2	H_2O
C-CO_2	87.9*	12	7.32	1.0	7.32	7.32			
H_2	10.3*	2	5.15	0.5	2.58				5.15
S	1.2*	32	0.04	1.0	0.04	0.04			
O_2(Ded)	0.5*	32	0.02	1.0	(.02)				
N_2	0.1*	28	0.00					0.00	
CO_2									
H_2O		18							
	100.0 lbs		12.53		9.92				

Theor. O_2 - 9.92
O_2 Excess - 1.19 (12% of 9.92)   1.19
Total O_2 - 11.11
N_2 (3.76 × O_2) - 41.77   41.77
Dry Air - 52.88
H_2O in Air - 0.536 (0.01015 Moles/Mole)   0.536
Air incl. H_2O - 53.416   7.36 | 1.19 | 41.77 | 5.686

Dry Flue Gas Heat Loss —

	Mcp		Moles		(T-t)		
CO_2 —	*10.0	×	(7.32)	×	420	=	30,744
SO_2 —	*10.4	×	(0.04)	×	420	=	175
O_2 —	* 7.25	×	(1.9)	×	420	=	3,624
N_2 —	* 7.05	×	(41.77)	×	420	=	123,681

158,224 Btu

Heat Loss from H_2O in Comb. Air —
    * 8.2 × (0.536) × 420   =   1,846 Btu

Heat Loss from H_2O in Fuel
           Moles
    * 8.2 × (5.15) × 420   =   17,737 Btu

Latent Heat Loss — H_2O in Fuel + H_2O formed from H_2
               Moles
   1040 × 18 × 5.15   =   96,408 Btu

Total Losses   =   274,215 Btu
* Total Input 100 × 18,500 Btu/lb   =   1,850,000 Btu
    % Heat Loss   —   14.82%
    % Comb Eff. 100 — 14.82   —   85.18%
Note—Boiler Eff = Comb Eff. — Radiation and Unaccounted for loss

Figure 6-6 Worksheet for "Mole" Method of Boiler Efficiency Calculation

methods require knowledge of the fuel analysis and the flue gas analysis. A worksheet for using the "mole" method with a fuel unit of 100 lbs of fuel is shown in Figure 6-6. This worksheet has been completed for a specific analysis by weight of a particular fuel oil but can be used for any fuel for which the analysis by weight is known.

In column 1 the percentage by weight for the fuel components is entered. Column 2 shows the molecular weights of the components. Column 3 is column 1 divided by column 2. Column 4 is the relationship between the moles of oxygen and the moles of the constituent in the particular chemical formula involved. Column 5 is column 3 multiplied by column 4. Column 5 divided by column 4 yields the values in the four right-hand columns.

The excess air (12 percent in this example) is determined from flue gas analysis. The moles per mole of water vapor in the combustion air are determined from the relative humidity and temperature of the combustion air. The mean specific heat (M_{cp}) can be determined for each flue gas constituent from the curves in Figure 6-7. With the tabulation shown, the first 5 of the 8 losses listed in Figure 6-3 are determined.

Heat loss from carbon in the refuse is calculated from the weight of ash per fuel unit and the percent carbon in the refuse. This determines the weight of carbon per fuel unit, which is then multiplied by 14,100, a commonly used value for the heating value of carbon.

For determining the loss for unburned combustible gas per fuel unit, the particular unburned gas and its percentage by volume in the flue gases must be known. If it is CO, the percent by volume as analyzed in the flue gas multiplied by the total number of moles in column 3 provides the number of moles of CO per fuel unit. This value multiplied by 28 (molecular weight of CO) provides the weight (lbs of CO). This weight multiplied by 4347 (the gross and also net heating value of CO) gives the unburned gas loss.

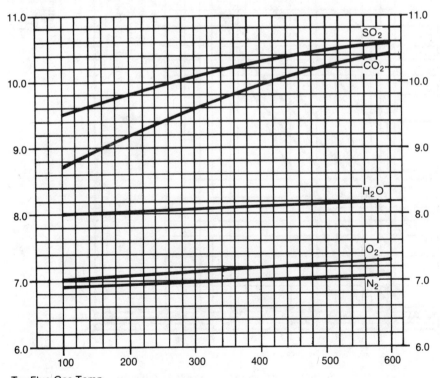

T = Flue Gas Temp.
Chart Shows Mean Molal Specific Heat between
Temperature T and Air Temperature, t = 60°

Figure 6-7 Mean Molal Specific Heat (M_{cp})

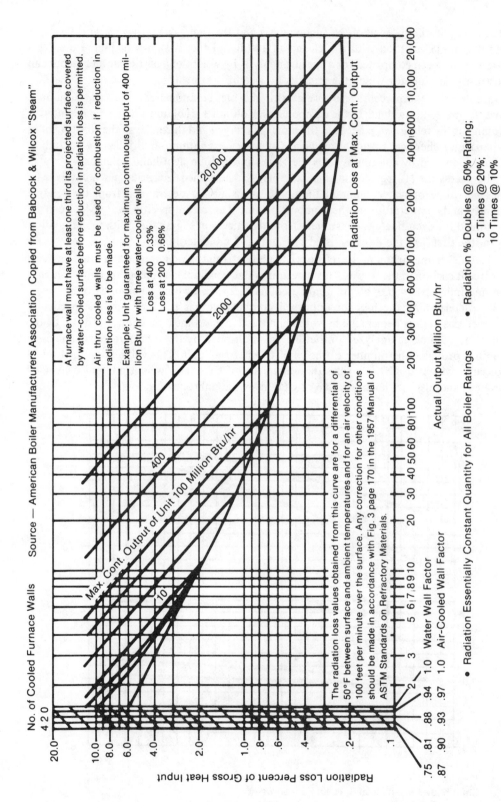

Figure 6-8 Radiation Loss

For a boiler fired with solid fuel, an unaccounted for loss of 1.5 percent is commonly used. For a gaseous or liquid fuel boiler, the commonly used value is 1 percent. The radiation loss is taken from the standard ABMA (American Boiler Manufacturers Association) curve shown in the upper part of Figure 6-8. Note that the percent radiation loss increases as the boiler is operated at reduced load. An analysis of these percentage values demonstrates that the heat loss from radiation for a particular boiler is essentially constant regardless of the boiler rating at the time.

The heat loss method involves a number of measurements and a much more involved calculation than the input/output method. Because of the manner in which measurements are used in the calculations, measurement errors do not cascade as in the input/output calculation method. Since the measurements used affect only the losses, a comparatively low percentage of the total measurement errors have a much smaller effect on the resulting boiler efficiency.

From a practical basis, a standard fuel analysis can usually be used without affecting the boiler efficiency more than approximately 1 percent. A practical method for quick use in the field is a simplified heat loss method that is based on curve fits of precalculated efficiency tables for a standard fuel. The fuel analysis and the precalculated tables are given in Figures 6-9, 10, 11, and 12. This shortcut method is shown in Figure 6-13. If the fuel analysis is unknown, the result of this shortcut method may be as precise as the detailed calculation using a standard fuel analysis. The simplified method requires a minimum of information. The essentials are % oxygen, flue gas and combustion air temperatures, type of fuel, boiler size, and percent rating.

Fuel		Bituminous Coal % by Wt.	No. 6 Fuel Oil % by Wt.	Natural Gas % by Vol.	Commercial Propane % by Vol.
Carbon	C	73.6	86.0	—	—
Hydrogen	H_2	5.3	11.0	—	—
Carbon Monoxide	CO	—	—	—	—
Methane	CH_4	—	—	82.9	—
Ethane	C_2H_6	—	—	14.9	2.2
Propane	C_3H_8	—	—	—	97.3
Pentane	C_4H_{10}	—	—	—	0.5
Illuminants	C_2H_4	—	—	—	—
Oxygen	O_2	10.0	1.0	—	—
Nitrogen	N_2	1.7	0.2	2.2	—
Carbon Dioxide	CO_2	—	—	—	—
Sulphur	S	0.8	0.8	—	—
Moisture	H_2O	0.6	1.0	—	—
Ash	—	8.0	—	—	—
		100.0	100.0	100.0	100.0
Dry Products of Perfect Combustion — cft of CO_2 and N_2		125.8	168.5	9.4	21.7
Air Requirement for Perfect Combustion — cft of Air		129.4	178.8	10.4	23.7
Ultimate CO_2 % by Volume		18.4	16.1	12.1	13.7
Heat Content Btu		13,640 per lb	18,873 per lb	1,107 per cft	2,576 per cft

Reference Temp. — 60°F, dry air basis.

Figure 6-9 Fuel Analysis for Combustion Efficiency Charts

CO$_2$		12.1	11.5	11.0	10.4	9.8	9.2	8.7	8.1	7.5	6.9	6.4	5.8
Excess Air		0	4.5	9.5	15.1	21.3	28.3	36.2	45.0	55.6	67.8	82.2	99.3
Oxygen		0	1	2	3	4	5	6	7	8	9	10	11
	°F												
	300	85.6	85.4	85.2	85.0	84.7	84.5	84.2	83.9	83.5	83.0	82.4	81.7
	350	84.6	84.3	84.1	83.8	83.5	83.2	82.8	82.4	81.9	81.3	80.6	79.8
	400	83.5	83.2	82.9	82.6	82.2	81.8	81.4	80.9	80.3	79.6	78.8	77.8
	450	82.5	82.1	81.8	81.4	81.0	80.5	80.0	79.4	78.7	78.9	77.0	75.9
	500	81.4	81.0	80.6	80.2	79.7	79.1	78.6	77.9	77.1	76.2	75.2	73.9
	550	80.3	79.9	79.4	79.0	78.4	77.8	77.2	76.4	75.5	74.5	73.4	71.9
	600	79.2	78.7	78.2	77.7	77.1	76.4	75.7	74.9	73.9	72.8	71.5	69.9
	650	78.1	77.6	77.1	76.5	75.8	75.1	74.3	73.4	72.3	71.1	69.7	67.9
	700	77.0	76.5	75.9	75.3	74.5	73.7	72.9	71.9	70.7	69.4	67.8	65.9
	750	75.9	75.4	74.7	74.1	73.2	72.4	71.5	70.4	69.1	67.7	66.0	63.9
	800	74.8	74.2	73.5	72.8	71.9	71.0	70.0	68.8	67.5	65.9	64.1	61.9
	850	73.7	73.1	72.3	71.6	70.6	69.7	68.6	67.3	65.9	64.2	62.3	59.9
	Loss per Percent Combustibles												
		2.8	3.0	3.2	3.4	3.7	4.0	4.3	4.6	5.0	5.5	6.1	6.8

Figure 6-10 Combustion Efficiency Chart — Gas

CO$_2$		16.1	15.3	14.5	13.8	13.0	12.2	11.5	10.7	9.9	9.2	8.4	7.7
Excess Air		0	4.7	9.9	15.7	22.2	29.5	37.7	47.1	58.0	70.7	85.7	103
Oxygen		0	1	2	3	4	5	6	7	8	9	10	11
	°F												
	300	89.7	89.5	89.2	89.0	88.8	88.5	88.1	87.8	87.3	86.7	86.1	85.4
	350	88.7	88.4	88.1	87.9	87.3	87.2	86.8	86.4	85.8	85.1	84.4	83.5
	400	87.6	87.3	87.0	86.7	86.3	85.9	85.4	84.9	84.2	83.4	82.6	81.5
	450	86.6	86.2	85.9	85.6	85.1	84.7	84.1	83.5	82.7	81.8	80.8	79.6
	500	85.5	85.2	84.8	84.4	83.8	83.3	82.7	82.0	81.1	80.1	79.0	77.6
	550	84.5	84.1	83.7	83.2	82.6	82.0	81.3	80.6	79.6	78.4	77.2	75.6
	600	83.4	83.0	82.5	82.0	81.3	80.7	79.9	79.1	78.0	76.7	75.4	73.6
	650	82.4	81.9	81.4	80.8	80.1	79.4	78.5	77.7	76.5	75.1	73.6	71.7
	700	81.3	80.8	80.2	79.6	78.8	78.1	77.1	76.2	74.9	73.4	71.8	69.4
	750	80.3	79.7	78.1	78.4	77.6	76.8	75.7	74.7	73.3	71.6	70.0	67.7
	800	79.2	78.6	77.9	77.2	76.3	75.4	74.3	73.2	71.7	70.0	68.1	65.7
	850	78.1	77.5	76.7	76.0	75.0	74.1	72.9	71.7	70.1	68.3	66.2	63.7
	Loss per Percent Combustibles												
		2.9	3.0	3.2	3.4	3.6	3.8	4.1	4.4	4.7	5.0	5.4	6.0

Figure 6-11 Combustion Efficiency Chart — No. 6 Oil

The curve in Figure 6-14 demonstrates the large influence of the radiation loss as boiler size and percent boiler rating are reduced. As shown on this example for a boiler of approximately 20,000 lbs/hr, an efficiency loss of 10 or more percentage points can occur if the boiler is operated over a 10 to 1 turndown range. The radiation loss is built into the boiler design and the user cannot operate the boiler in any manner to reduce this loss.

CO_2		18.4	17.5	16.6	15.7	14.9	14.0	13.1	12.2	11.4	10.5	9.6	8.7
Excess Air		0	4.9	10.2	16.2	22.9	30.4	38.8	48.5	59.5	72.9	88.4	107
Oxygen		0	1	2	3	4	5	6	7	8	9	10	11
	°F												
	300	91.5	91.3	91.1	90.8	90.5	90.2	89.9	89.5	89.0	88.5	87.8	87.1
	325	91.0	90.8	90.6	90.3	89.9	89.6	89.3	88.8	88.3	87.7	86.9	86.1
	350	90.5	90.3	90.0	89.7	89.3	89.0	88.6	88.1	87.5	86.8	86.0	85.1
	375	90.0	89.8	89.5	89.1	88.7	88.4	87.9	87.4	86.7	86.0	85.1	84.1
	400	89.5	89.2	88.9	88.5	88.1	87.7	87.2	86.6	85.9	85.1	84.2	83.1
	425	89.0	88.7	88.4	88.0	87.5	87.0	86.5	85.9	85.2	84.3	83.3	82.1
	450	88.5	88.2	87.8	87.4	86.9	86.4	85.9	85.2	84.4	83.4	82.4	81.1
	475	88.0	87.7	87.3	86.8	86.4	85.7	85.2	84.5	83.7	82.6	81.5	80.1
	500	87.5	87.1	86.7	86.2	85.7	85.1	84.5	83.7	82.9	81.8	80.6	79.1
	525	87.0	86.6	86.2	85.6	85.0	84.4	83.8	83.0	82.1	80.9	79.7	78.1
	550	86.5	86.1	85.6	85.0	84.4	83.8	83.1	82.2	81.3	80.0	78.7	77.1
	575	86.0	85.5	85.0	84.4	83.9	83.1	82.4	81.5	80.5	79.2	77.8	76.1
	600	85.4	85.0	84.4	83.8	83.2	82.5	81.7	80.7	79.7	78.3	76.8	75.1
	625	84.9	84.5	83.9	83.3	82.6	81.8	81.0	80.0	78.9	77.5	75.9	74.1
	650	84.4	83.9	83.3	82.7	82.0	81.1	80.3	79.3	78.1	76.6	75.0	73.1
	675	83.9	83.4	82.8	82.1	81.4	80.5	79.6	78.6	77.3	75.8	74.1	72.1
	700	83.3	82.8	82.2	81.5	80.7	79.8	78.9	77.8	76.5	74.9	73.2	71.1
	725	82.8	82.3	81.7	80.9	80.1	79.2	78.2	77.1	75.7	74.1	72.3	70.1
	750	82.3	81.7	81.1	80.3	79.5	78.5	77.5	76.3	74.9	73.2	71.3	69.1
	775	81.8	81.2	80.5	79.7	78.9	77.9	76.8	75.6	74.1	72.4	70.4	68.1
	800	81.2	80.6	79.9	79.1	78.2	77.2	76.1	74.8	73.3	71.5	69.4	67.0
	825	80.7	80.1	79.4	78.5	77.6	76.6	75.4	74.1	72.5	70.7	68.5	66.0
	850	80.1	79.5	78.8	77.9	77.0	75.9	74.7	73.3	71.7	69.8	67.6	65.0
	875	79.6	79.0	78.2	77.3	76.4	75.3	74.0	72.6	70.9	69.0	66.7	64.0
	Loss per Percent Combustibles												
		3.0	3.1	3.3	3.5	3.7	3.9	4.2	4.5	4.8	5.2	5.7	6.3

Figure 6-12 Combustion Efficiency Chart — Coal

(As used in Figures 6-10, 6-11, and 6-12, combustion efficiency is boiler efficiency before radiation loss and unaccounted for loss are deduced.)

% Sensible Heat Loss (all flue gas)
$\Delta t \times [0.023 + 0.00011\ (\%\ O_2 + 1)^2]\,$*

% Latent Heat Loss

Natural gas, 9%

No. 2 oil, 5.5%

No. 6 oil, 5.0%

Coal, $[2.7 + 12.8\ (M^2 + M)]\%$
    M equals fuel moisture % as a decimal number

% Radiation Loss	20,000 lb/hr blr	100,000 lb/hr blr
Full load	1.0%	.4%
50% load	2.0%	.8%
25% load	4.0%	1.6%

% Boiler Efficiency

100 – % Sensible Heat Loss –
      % Latent Heat Loss –
      % Radiation Loss

* Δt, Flue Gas Temperature – Combustion Air Temperature

Figure 6-13 Heat Loss Method Shortcut

(S. G. Dukelow Formula — completely empirical method)

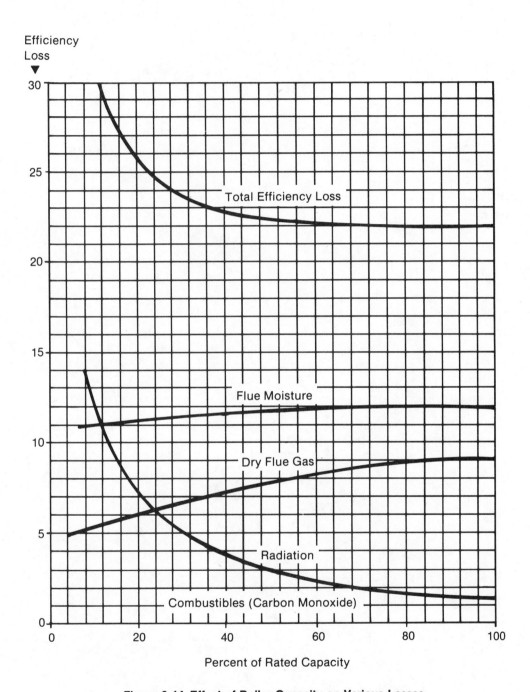

Figure 6-14 Effect of Boiler Capacity on Various Losses

Section 7
The Steam Supply System and Firing Rate Demand Control Loops

7-1 Saturated Steam Moisture Elimination

In Section 2 the water circulation of watertube boilers was explained. The steam that is collected in the steam drum contains some droplets of boiler water that must be removed before the steam is delivered from the drum. Figure 7-1 demonstrates in a general way how the water and steam are separated.

As the steam-water mixture rises to the drum, internal baffles separate this mixture from the water that is entering the downcomer tubes. The steam is liberated from the mixture and collects in the upper part of the drum. Before passing out of the drum, the steam passes through mechanical separation devices. These "scrubbers" or "separators", which may be of many different designs, return the moisture droplets to the water in the drum and allow dry steam to pass out of the drum.

The most sophisticated of these devices uses a centrifugal action that whirls the mixture with the steam emitted from the center and the heavier water from the outside. When very dry steam is required, there will be many of these devices inside the drum along with simpler scrubber equipment. While it is not evident from the outside, the inside of the drum is often filled with such devices. A typical arrangement is shown in Figure 7-2.

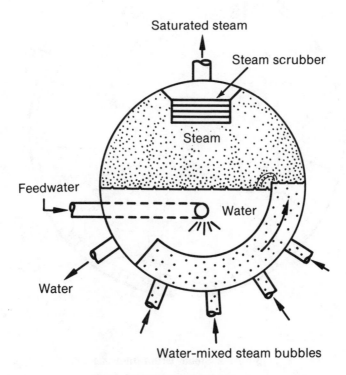

Figure 7-1 Boiler Steam Drum

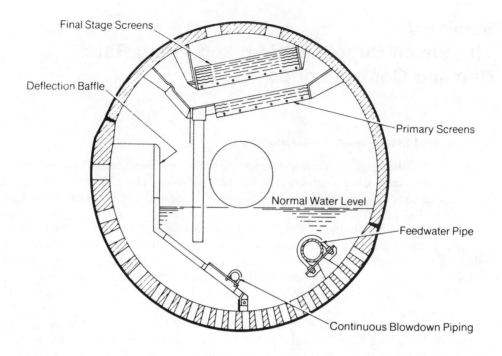

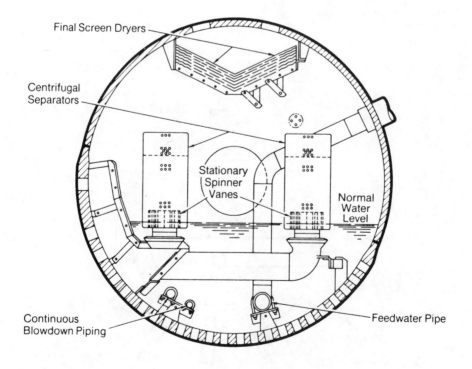

Figure 7-2 Steam Drum Internals

(From *Fossil Power Systems*, ©Combustion Engineering Co., Inc., 1981.)

7-2 Superheated Steam Temperature vs Boiler Load

Unless a boiler making superheated steam is equipped with a control mechanism, the temperature of the superheated steam will vary, depending on various operating factors. Most industrial boilers that generate superheated steam do not include control mechanisms. Operating factors that may cause the temperature of superheated steam to vary are boiler load, steam pressure, excess combustion air, specific fuel burned, cleanliness of the heat transfer surface, and others.

In the characteristic of steam temperature vs boiler load, the design of the superheater is particularly significant. The steam leaves the drum and enters the superheater. If this heat exchanger is located so that it can directly "see" the flame, it receives radiant heat from the flame and is called a radiant superheater. If the location is such that the superheater cannot "see" the flame and receives all its heat by convection, it is called a convection superheater.

The opposite characteristics of steam temperature vs firing rate for these two different types of superheaters are shown in Figure 7-3. Note that with a radiant superheater the temperature rises as steam output is decreased. This results from a relatively constant radiant input with a reduction of the "cooling" effect of the steam as load is decreased. With a convection superheater the opposite is true. The heat transferred rises as steam flow increases, but the rate of temperature increase becomes less as steam flow continues to increase.

By combining radiant and convection superheater elements in series, steam temperature changes a smaller amount as the steam load changes. This method has been used considerably by one boiler company in particular to achieve a more constant steam temperature. The advantage results from the more constant temperature that is obtained without the complexity and expense of resorting to control mechanisms.

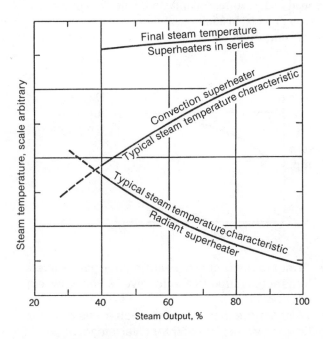

A substantially uniform final steam temperature over a range of output can be attained by a series arrangement of radiant and convection superheater components.

Figure 7-3 Superheater Characteristics

(From *Steam, Its Generation and Use,* ©Babcock & Wilcox)

When temperature is not controlled the normal practice is to design for maximum temperature at full load, as shown in Figure 7-4, with temperature reducing as load is reduced. Since no control is involved, steam temperature is affected by the operational factors mentioned above. The type of fuel changes the flame temperature, producing a change in furnace heat absorption and thus a change in the temperature of flue gases entering the superheater. The cleanliness of the heat transfer surfaces changes their heat transfer coefficient, resulting in a change in heat absorption. A change in the amount of excess combustion air changes flue gas mass flow, resulting in a change in the heat transfer coefficient; it also changes furnace outlet flue gas temperature.

7-3 Mechanisms for Control of Superheat Temperature

Typical industrial boilers designed for the control of steam temperature may be capable of achieving full design steam temperature at 30 percent to 40 percent of full load steam output. Above this load point, a control mechanism is used to reduce the steam temperature as the load is increased. The result is that there is a possible family of curves of steam temperature vs load for such a boiler, as shown in Figure 7-5.

The primary purpose of the control mechanism is to adjust the superheating capacity as steam output changes. The boiler steam temperature is also affected, however, by the cleanliness factor, the fuel being fired, the imbalance between fuel Btu input and steam Btu output, and excess combustion air. The control mechanism must also have the capability of adjusting for these secondary influences in order that the boiler may be controlled at a constant steam temperature.

The purpose of steam temperature control is to obtain as nearly as possible a constant superheat temperature at all boiler loads. The primary benefit in constant steam temperature is in improving the economy of conversion of heat to mechanical power. Control capability increases the lower load temperature, resulting in the potential for higher thermal efficiency of the power generation process. In addition, maintaining a constant temperature minimizes unequal expansion or contraction due to unequal mass or material of the various rotating

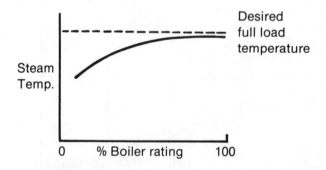

- Temperature increases with higher gas temperatures across superheater for given steam flow (e.g., dirty boiler).
- Temperature increases with higher flue gas mass flow across superheater for given steam flow (e.g., higher excess air).
- Temperature entering superheater is a function of furnace temperature and furnace heat absorbed.

Figure 7-4 Uncontrolled Superheat (Typical)

parts of power generation machines. This makes possible the use of smaller clearances and results in higher thermal efficiency in the energy conversion process.

The control mechanisms that are operated by the steam temperature control equipment may involve the fire side or the water side of the boiler. The basic fire side mechanisms change either the temperature of the flue gases entering the superheater or the mass of the flue gases entering the superheater, or both. Different boiler manufacturers may use different methods.

Figure 7-6A demonstrates a method of changing the temperature of the flue gases entering the superheater. In this method, used by Combustion Engineering, Inc., burners mounted in the furnace corners are arranged so that the flame can be tilted up or down from horizontal. The burner flame is aimed at a tangent to an imaginary circle in the center of the furnace, and the burners in the four corners of the furnace are all tilted at the same angle. The result is a "fireball" in the center of the furnace, which rotates and which can be raised or lowered in the furnace by changing the tilt of the burners. Lowering the fireball increases furnace heat absorption, which lowers the flue gas temperature as it enters the superheater. Raising the fireball decreases the furnace heat absorption with the opposite effect on the temperature of the flue gases entering the superheater.

Figure 7-6B shows a furnace with burners mounted in fixed furnace wall positions. Note that the burners are mounted at higher and lower elevations. By varying the ratio of the fuel fired in the upper row of burners to that in the lower row of burners, the furnace heat absorption can be modified, thereby changing the temperature of the flue gases entering the superheater.

The flow stream of the flue gas passing the superheater can be split so that the mass of flue gas in contact with the superheater can be varied. Such a mechanism is called a superheater bypass damper and is shown in Figure 7-6C. The opposite of this is shown in Figure 7-6D. In this mechanism a flue gas recirculating fan adds a variable amount of flue gas mass flow to the stream in contact with the superheater. A further fire side method is the raising and lowering of the percentage of excess air in order to control steam temperature. This method tends to improve the overall heat rate of power generation equipment though the thermal efficiency of the boiler itself may be lowered.

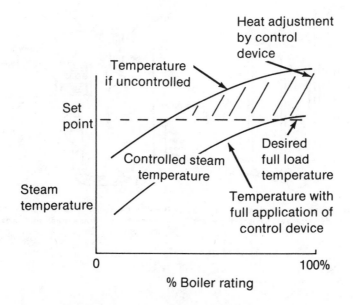

Figure 7-5 Controlled Superheat (Typical)

In addition to the fire side methods described above, there are three basic types of water side methods. Figure 7-7A shows the use of a spray mechanism to spray water into the superheated steam. Varying the water flow raises or lowers the steam temperature. Figure 7-7B demonstrates a mechanism using a control valve to divert part of the steam to a shell-and-tube heat exchanger. The steam is cooled in the heat exchanger and then mixed again with the rest of the steam, raising or lowering its temperature. The heat exchanger is located in either the steam drum or the mud drum. Figure 7-7C uses a shell-and-tube heat exchanger in the saturated steam line between the boiler and the superheater. A controlled portion of the feedwater to the boiler is diverted to the heat exchanger to remove a variable amount of the latent heat, thus raising or lowering the final steam temperature.

These various methods can be used singly or in combination to control the final steam temperature. Which method or combination of methods is used depends upon a number of factors. The particular boiler manufacturer and that company's design philosophy and best competitive offering are important considerations in their selection of the control means and its application.

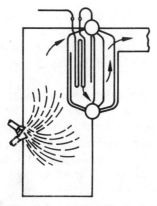

(A) Tilting burners varies super heat. Heating is lowered by directing flame downward; raising increases superheater temperature.

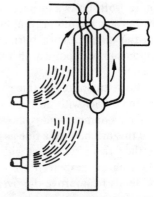

(B) Lighting additional burners raises steam temperature, increases superheater action; upper burner row is most effective.

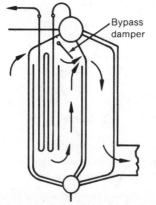

(C) Dampers bypass portion of combustion gases around superheater and reheater; are most effective in upper steaming range.

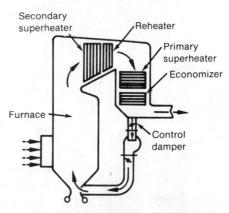

(D) Partial gas recirculation affects the overall furnace temperature and, as a result, influences the total heat absorption.

From *Power* magazine Special Report, "Steam Generation," by Rene J. Bender, Associate Editor, McGraw Hill, NY.

Figure 7-6 Fire Side Control Mechanisms

From a control standpoint, the strategy must be based on the particular mechanisms used and the manufacturer's philosophy for controlling steam temperature. The control characteristics of the different methods may be quite different. The time constant for this process is in minutes. The most rapid response is from spray water, and the slowest response is normally the method that extracts a part of the latent heat as shown in Figure 7-7C. Another characteristic of this process from a control standpoint is that response time is often a variable function of the steam flow rate.

7-4 Steam Temperature Control Strategy

The strategy used for control of steam temperature for any particular boiler is normally recommended by the manufacturer of that boiler. In a few installations with the steam flow rate reasonably constant, a single-element feedback system may operate satisfactorily. In the typical installation, some form of feedforward control, cascade control, or a combination of these is required. The normal control requirement is to control the temperature within plus or minus 10° F. Figures 7-8 and 7-9 demonstrate two methods of controlling superheat temperature using a water spray as shown in Figure 7-7A.

Figure 7-8 shows the application of a feedforward-plus-feedback strategy. Since air flow rate is an index of firing rate and excess combustion air, the air flow measurement is used as the anticipatory or feedforward signal. In the summer (x), this signal is combined with the output of the signal from the steam temperature feedback controller (y). The output of this

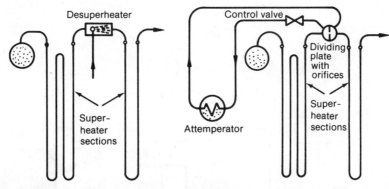

(A) Pure water sprayed into the superheated steam vaporizes; degree of superheat is reduced by heat of vaporization.

(B) In an effort to avoid steam contamination, desuperheating can also be done through use of shell-and-tube heat eachanger.

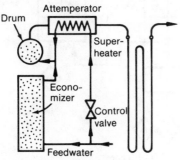

(C) In this hookup the saturated boiler steam is partially condensed by feedwater, controlling the steam temperature

From *Power* magazine Special Report, "Steam Generation," by Rene J. Bender, Associate Editor, McGraw Hill, NY.

Figure 7-7 Water Side Control Mechanisms

summer provides a signal for the spray water flow control valve. Note that the feedback controller is supplied with an override controller (w).

Controller (w) is a form of override controller that provides a minimum output value tracking signal for controller (y). This is included so that when the boiler load is below that of the steam temperature control range, the output of controller (y) will be the signal necessary so that the output signal of the summer (x) will provide a "just closed" position of control valve (u). This function is necessary for good control, since on increasing or decreasing steam flow rates, the steam temperature may be at the design temperature level with different firing and air flow rates. This relationship is also affected by the rate of load change.

The proportional (t) is shown to indicate that scaling of the air flow signal is necessary. In practice this block would probably not exist since input scaling capability is an integral part of most summer hardware or software blocks. The bias logic (v) is provided to obtain a positive value signal of the output of summer (x) when the signal to the control valve is reduced to 0 percent. As shown this is a 5 percent bias. This allows a set point of the override controller to be a value of 5 percent with control action above and below this signal level. The f(x) logic (z) allows for a nonlinear relationship between the measured air flow signal and the position demand signal to the spray water control valve.

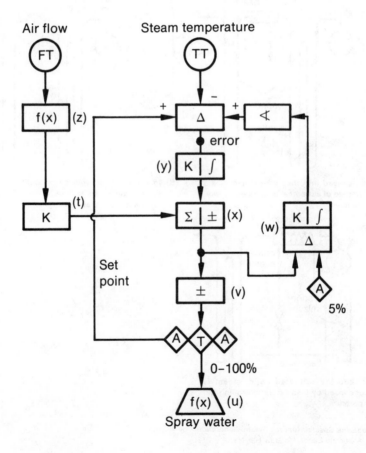

Figure 7-8 Feedforward-plus-Feedback Control of Superheat Spray

In calibrating and tuning this system, the relationship between air flow rate and spray water control valve input signal is determined by steady state testing at the design steam temperature and while operating at different boiler steam flow rates. When this relationship is known, the air flow signal is scaled and functionally adjusted by modifying the *f(x)* function and by changing the input gain of the summer (x).

With the air flow signal at the level where full steam temperature is obtained without spray water, and with the input gain of that signal adjusted, the second summer signal is adjusted to a 50 percent value with a summer input gain of 1.0. The output bias of summer (x) is adjusted so that the summer output will be at a plus 5 percent value and bias (v) adjusted to a minus 5 percent value. The output of the bias (v) will then be a 0 percent signal to the control valve.

An examination of the loop of controller (w) shows that, since there is practically no process time constant, its action will be very fast. The gain and integral settings of this controller can usually be arbitrarily set at high values with precise tuning unnecessary. With the system calibrated in this manner, on line tuning of the steam temperature controller (y) is the only remaining action. Preliminary tuning should be under steady state operation followed by testing under dynamic boiler load conditions. Small tuning adjustments may be necessary due to imperfect or variable relationships in the feedforward portion of the system.

The above description demonstrates that a multivariable system must be calibrated in addition to the normal proportional, integral, and derivative tuning of the measured variable controller. This calibration should always be done before any tuning of the controller. The description also demonstrates the necessity for the control designer to write a complete functional description of the control system to communicate the system features and the designers intent. The SAMA diagram displays the functions of the system but does not explain why they have been included by the control system designer.

This system relies on a predictable relationship between the spray water control valve position and spray water flow. If this is not the case then a cascade spray water flow control should be added. The valve position demand signal would then become the set point of a spray water flow controller.

Figure 7-9 shows an alternate cascade control. In this case the primary control is a feedback controller (a,b,c) from final steam temperature. The control logic of this controller is split into three parts to allow the tracking logic to be implemented. Transfer switch (h) is open when both logic switches (f) and (g) are closed. This occurs when the final steam temperature is below the set point and the spray water valve is closed, indicating that the boiler load is below that of the steam teamperature control range.

With switch (h) open, the integral action of controller (c) is stopped and the output of controller (c) tracks the signal from the steam temperature measurement downstream of the spray nozzle. This keeps controller (a,b,c) ready to immediately assume control when the boiler load again enters the range in which the spray water can be used to control steam temperature.

The secondary controller (d) is a much faster control loop that uses feedback from steam temperature immediately downstream from the spray nozzle. As in the feedforward arrangement, the anti-windup feature is needed when the boiler load is such that no spray water is required.

In tuning the above system, as in any cascade control loop, the secondary controller which operates the spray control valve is tuned first. The gain and integral of controllers (a) and (c) are set at very low values to stabilize the set point input to controller (d). The boiler is then operated at a stable steam flow rate within the steam temperature control range.

The gain and integral settings of controller (d) are adjusted to low values and with a resulting steady steam temperature at the spray nozzle outlet. The gain is then increased until the control action becomes unstable. The gain is then reduced as the integral setting is increased until the optimum control pattern for the temperature downstream from the spray nozzle is obtained. Analytical techniques as described in the literature or tuning aid devices can also be used. The primary controller (a,b,c) from final steam temperature is then tuned in a similar fashion.

7-5 Steam Supply Systems

A generic steam supply system is shown in the diagram in Figure 7-10. This diagram shows two saturated steam boilers connected to a steam header. The normal measurement points of a flow nozzle or orifice plate for measuring steam flow and a connection for steam pressure measurement are shown. If these were superheated steam boilers, a steam temperature measurement would also be shown. At the outlet of each boiler is a non-return valve that prevents steam from the header from entering the steam drum. The steam header is the collecting point for the steam from more than one boiler. While two boilers are shown here, there could be multiple boilers operating at the same pressure and feeding steam to the same header.

The pressure in the steam header is also normally measured. On the outlet side of the steam header, the steam may travel through many steam lines in its transit to the users of the steam. The basic need of the steam user is for heat. Steam is a convenient carrier of that heat. When the header, the boiler steam leads, and all the steam lines are pressured to the normal operation pressure, a certain quantity of steam and therefore heat is stored in the system.

7-6 Heat Energy and Water Storage

Figure 7-11 is a diagram typical of the heat storage in the boiler. Heat is stored in the water, steam, metal, refractory material, and insulation material of the boiler. When the

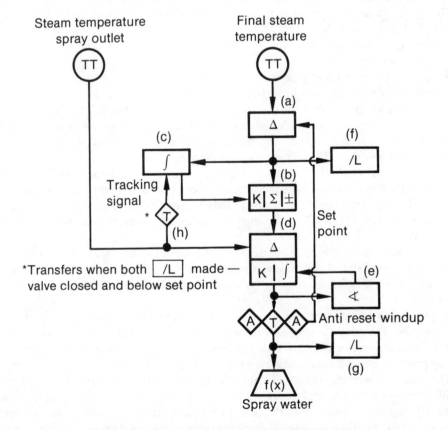

Figure 7-9 Cascade Control of Superheat Spray

boiler is at normal pressure and corresponding temperature but with no steam output, the diagram shows that a large percentage of the heat storage at maximum boiler load already exists. Similarly all the boiler steam space is filled with steam, which represents heat storage. In this case there are no steam bubbles underneath the surface of the boiler water since no steam is being produced. Heat is also stored in the boiler water, and since there are no steam bubbles, the entire water space below the normal level in the boiler drum is water at saturation pressure and temperature. At this time the boiler is also losing heat by radiation.

As additional fuel and air are burned, the boiler steam output may be increased to 100 percent or more of its rated value. Note that as firing rate increases, the heat energy storage in metal, refractory, etc., increases due to the higher operating temperature of these parts. The steam space of the boiler is increased due to the steam bubbles below the steam drum water surface, and this increases the energy storage in the boiler steam. The steam bubbles subtract from the water volume and this subtracts from the boiler water energy storage as boiler steam flow is increased. Another aspect of the reduction in boiler water volume is that the mass of water in the boiler is less at higher boiler steam flow rates. This is an important factor in the proper control of boiler feedwater.

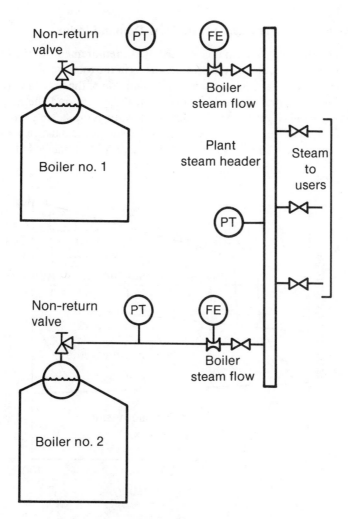

Figure 7-10 Steam Header for Multiple Boilers

The energy storage in the boiler acts the same as a flywheel, which is also an energy storage device. Table 7-1 can be used for demonstration.

Table 7-1
Energy Storage Relationships

Water at 200 psig sat.	355.5 Btu/lb
Latent heat of evap.	842.8 Btu/lb
Water at 180 psig sat.	346.2 Btu/lb
Heat released by dropping pressure (200 to 180 psig)	9.3 Btu/lb

As pressure is reduced, the water cannot exist at 355.5 Btu/lb. It immediately releases 9.3 Btu/lb, thus cooling the water to the saturation temperature for 180 psig. The heat released causes an immediate generation of steam that carries a corresponding amount of heat. In this way heat is released from storage as the pressure drops and must be replaced in storage as the pressure increases.

7-7 Steam Pressure — The Basic Steam Flow Demand Index

The demand on the boiler system is generated by the requirements of the steam users. As they open valves to get more steam, the pressure drops in the total storage system, triggering the release of some of the heat from storage. The magnitude of the pressure drop depends on the relationship between the total volume of the steam system and the magnitude of the change in steam demand. If the volume is relatively low, then the steam pressure change will be relatively high and vice versa. The steam header pressure is the balance point between the demands of the steam users and the supply of fuel and air to the boilers. At a constant steam

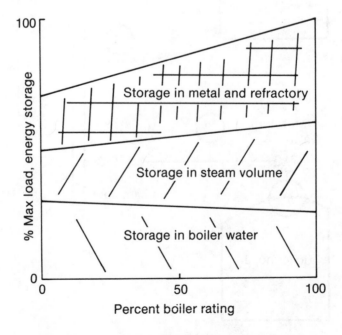

Figure 7-11 Characteristics of Boiler Energy Storage

flow requirement, a constant pressure in the steam header indicates that energy supply and demand are in balance. The balance is represented by the statements that follow.

(1) Steam demand = steam flow plus or minus (K) $\times$ (pressure error). K is a function of the system volume related to the demand flow rate.

(2) Supply side = fuel, air, and water energy to the boiler plus the change in energy storage.

(3) Demand side = steam to users.

(4) Balance point = steam header pressure.

(5) Pressure at set point — demand equals supply when energy storage is constant.

6) Pressure increasing — supply exceeds demand (may equal demand if storage is decreasing)

(7) Pressure decreasing — demand exceeds supply (may equal supply if storage is increasing)

7-8 Linking the Steam Pressure Change to Changes in Firing Rate

The combustion, feedwater control, and steam temperature control systems determine how a boiler actually operates and whether it achieves its efficiency potential. The controls should be designed to regulate the fuel, air, and water to a boiler and maintain a desired steam pressure or hot water temperature while simultaneously optimizing the boiler efficiency.

During either normal or abnormal operation, the greater the sophistication of the controls, the greater the efficiency potential of the boiler system. A control system can usually be upgraded in its functions by adding additional components or software. Improving a control system is usually a cost-effective way to improve the operating efficiency of any boiler.

Generally the controls are classified in two main groups: on/off and modulating. On/off controls are subdivided into basic on/off (full on and off) and high/low/off, which has a high and low fire "on" condition plus the "off" condition. Modulating controls are subdivided into two basic classes: positioning and metering.

The simplest, most basic, and least costly control and the one used to control firing rate on only the smaller firetube and watertube boilers is on/off. The control is initiated by a steam pressure or hot water temperature switch. As the pressure or temperature drops to the switch setting, the gas valve is opened (or the fuel pump started) along with the combustion air fan motor. The fire is ignited usually with a continuous pilot flame. The fuel and air continue operating at full firing rate capacity, and the pressure or temperature rises until the switch contact is opened. Figure 7-12 represents an on/off system.

Although such a system may maintain steam pressure or hot water temperature within acceptable limits, combustion is not controlled because combustion efficiency (while firing) is a result of mechanical burner adjustment. When the burner is on, the excess air is subject to the following variations in the fuel supply.

- Pressure and temperature of the fuel
- Btu content of the fuel (hydrogen/carbon ratio)
- Fuel specific gravity
- Fuel viscosity
- Mechanical adjustment tolerances

The excess air is also subject to variations in the combustion air supplied.

- Air temperature and relative humidity
- Air supply pressure
- Barometric pressure
- Mechanical adjustment tolerances

In addition, each time the burner is off, unless the flue damper is closed, cold air passes through the boiler carrying heat up the stack. Using this system, the "on" fire is at full firing rate and the flue gas temperature is at maximum. Figure 7-13 represents how the on/off system works.

The other on/off system is the high/low/off control in which the burner system has two firing rates called "high fire" and "low fire". If the Btu requirements are between those of high fire and low fire, the burner will stay on all the time, cycling between high and low. Unless the load is below low fire input, this eliminates the "off" heat losses caused by cold air through the boiler. Such a system has three steam pressure or hot water temperature settings.

(1) Stop fire or off
(2) Start boiler and go to low fire or stop high fire and go to low fire
(3) Start high fire

This system will hold the steam pressure or the hot water temperature within closer tolerances of the desired steam pressure or hot water temperature. It will have a lower weighted average flue gas temperature than a straight on/off system but a higher weighted average flue gas temperature than a fully modulating control.

The system can be tuned to burn the fuel efficiently when the burner is "on" at any time. It will get out of tune when any of the fuel and air conditions change from those present when the burner was set. Compromising some of the mechanical adjustments may be necessary in trying to optimize combustion at both the "low fire" and "high fire" settings of the high/low/off system.

Using on/off control to add water to a boiler — based on the water level in the boiler —intermittently cools and heats the boiler water, causing increased on/off or high/low cycling action of the firing rate control.

The action of the high/low/off control under the same load conditions as Figure 7-13 is shown in Figure 7-14.

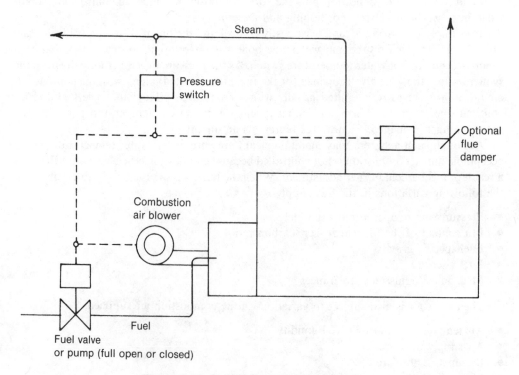

Figure 7-12 On/Off System Arrangement

Modulating control is a basic improvement in controlling combustion and feedwater. A continuous control signal is generated by a controller connected to the steam or hot water piping system. Reductions in steam pressure or hot water temperature increase the output signal, which calls for a proportionate increase in firing rate.

Modulating control is an improvement because the fuel and air Btu input requirements and the Btu of the steam or hot water output of the boiler are continuously matched. The action of such a control system under the same load conditions of Figure 7-13 and Figure 7-14 is shown in Figure 7-15.

Because matching the input and output Btu requirements is improved, the steam pressure or hot water temperature is maintained within closer tolerances than is possible with the previously discussed control systems. The weighted average flue gas temperature is lower, so boiler efficiency is greater. Table 7-2 compares the efficiency of boilers with the different systems while operating under the indicated load conditions. The influences of changes in the condition of fuel and air have been eliminated by assuming a 10 percent excess air and a constant flue gas temperature for each of the loads or firing rates that would occur.

Although each boiler will have its own characteristics of excess air as opposed to flue gas temperature, the table is typical for a gas-fired boiler with the efficiency calculations based 450 to 600 degrees flue gas temperature over the 25 percent to 100 percent load range. The table represents control systems operating ideally with no variation in excess air. Excess air should be higher at loads less than 50 percent, and efficiencies would be lower than those shown for the high/low/off and modulating systems.

At 25 percent load, high/low/off and modulating have the same efficiency. Twenty-five percent was considered low fire, which would be "on" full time, or the same continuous firing

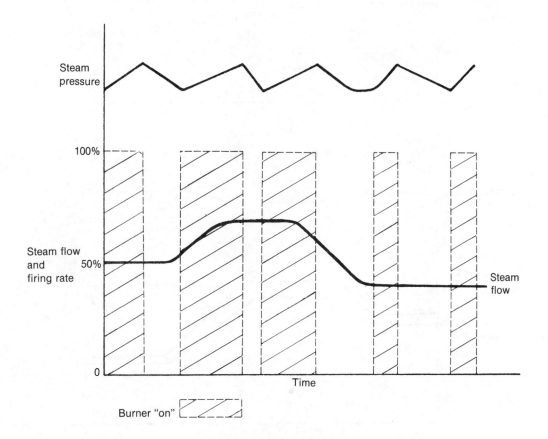

Figure 7-13 On/Off Control Action

Table 7-2
Control System Perfection Comparison

Type of Control	Efficiency at % load			
	25%	50%	75%	100%
On/off	70.28	74.28	75.61	76.28
On/off with flue damper	73.28	75.28	75.95	76.28
Hi/low/off	76.88	76.48	76.35	76.28
Modulating	76.88	77.68	77.15	76.28

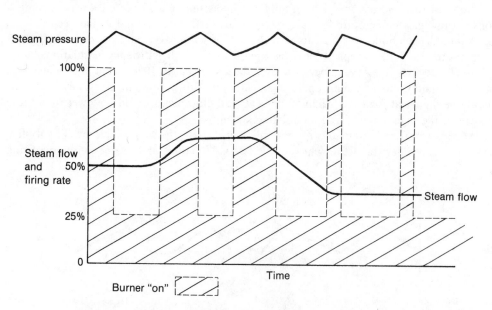

Figure 7-14 High/Low/Off Control Action

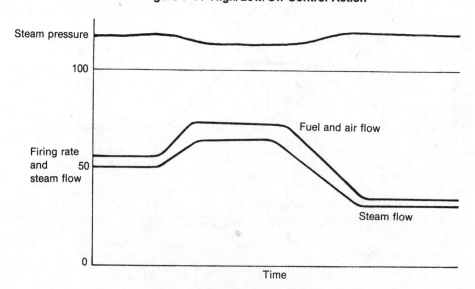

Figure 7-15 Modulating Control Action

rate for both systems. At 100 percent, all systems would be "on" full time at 100 percent firing rate. Therefore, results are the same for all systems. In the middle range, efficiency clearly improves as control sophistication is increased. The benefit of modulating control is clearly established by Table 7-2. The type of modulating control and how it is implemented in developing the "firing rate demand" signal is next examined.

7-9 Steam Pressure or Steam Flow Feedback Control

Assuming that the equipment for generating a firing rate demand signal is of the modulating type, there are several different methods and considerations involved. For the simpler systems, a simple proportional or proportional-plus-integral feedback controller may be used. Figure 7-16 demonstrates the normal method for regulating the firing rate using steam pressure. In some installations a constant steam flow may be required for one or more boilers in combination, while other boilers connected to the same header are used for controlling steam pressure. It is also possible to arrange the system as shown in Figure 7-17 so that the control for a particular boiler can be switched between steam pressure and steam flow control. The switching procedure would require the boiler operator to switch the control to manual, adjust the set point to the desired value of the variable being switched to, operate the transfer switch, and then transfer the control back to automatic operation.

Figure 7-18 is a diagram of a change in steam flow rate, firing rate, and steam pressure with respect to time. This diagram is useful in analyzing the tuning requirements of the loop. Suppose, as shown, the steam pressure transmitter has a range of 0 to 300 psig for a 0 to 100 percent output signal and that at a constant steam flow rate the firing rate is 90 percent of its range when steam flow is 100 percent of its range. If 10 psig is the minimum desired pressure deviation for a 10 percent change in steam flow rate, such a deviation will produce a (10/300) 0.033 or 3.3 percent change in the signal from the steam pressure transmitter.

This signal must be amplified to produce an immediate change in the firing rate. Under steady-state conditions the firing rate change for the 10 percent steam flow rate change is approximately 9 percent of its range.

A load increase, however, requires overfiring to add the required additional energy storage that will allow the pressure to return to its set point. The magnitude of the desired temporary overfiring is usually approximately 25 percent of the steady-state firing rate change, or 2.25 percent in this case. The 9 percent steady-state firing rate increase is thus increased to a needed 11.25 percent of maximum. Since the multiplier is (11.25/3.3) 3.41, a

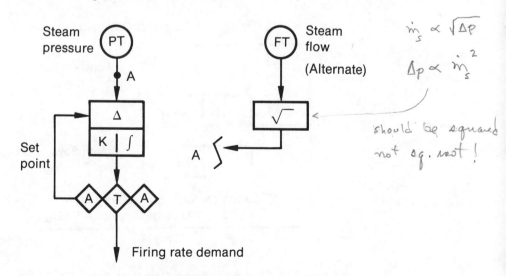

$$\dot{m}_s \propto \sqrt{\Delta p}$$

$$\Delta p \propto \dot{m}_s^2$$

should be squared not sq. root!

Figure 7-16 Steam Pressure or Steam Flow Feedback Control

gain of 3.41 would be applied to the steam pressure controller. Note that if the steam pressure transmitter had a maximum range of 0 to 400 psig, the deviation would have been 0.02 or 2 percent and the multiplier would have been (11.25/2) 5.625. Upon a reduction in steam flow, underfiring would have been necessary to adjust the system energy storage.

For tuning such a steam pressure controller, a typical controller gain of 4.5 is a reasonable starting point with an integral setting of 0.25 repeats per minute. As stated earlier, the optimum gain will be determined by the ratio of boiler capacity to steam system volume. The

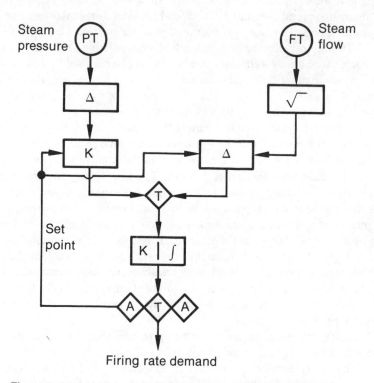

Figure 7-17 Steam Pressure or Steam Flow Feedback Control

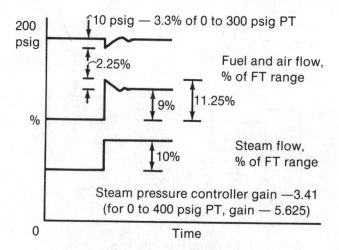

**Figure 7-18 Changes in Steam Flow Rate, Firing Rate,
and Steam Pressure with Respect to Time**

optimum integral setting will be determined by the time constant of the particular steam generation process. In this case, since the time constant is several minutes, the integral value in repeats per minute will be less than 1.0.

The optimum controller tuning may not be that which will produce the optimum steam pressure pattern. In many cases it is possible to obtain improved steam pressure control at the expense of boiler operating efficiency. Increasing the controller gain may produce oscillations of the fuel flow and the air flow that may improve steam pressure control while decreasing efficiency due to the oscillations. The degree to which this may occur is different for different installations. The resolution of this question must be based on the judgment of the process experienced individual who is responsible for the controller tuning.

If steam flow is the controlled variable, the gain of the controller will approximate that of a flow control loop and will probably be less than 1.0. Since the integral value is related to the process time constant, it will approximate the value of the pressure control integral value.

The output of the controller is called firing rate demand. Generation of a proper firing rate demand signal is of primary importance since all control of fuel and air is directed from this master signal. The controller is therefore called the master controller. Because of the importance of this controller, there are a number of more sophisticated configurations available. The use of these more complex arrangements results in greater precision of the firing rate demand control signal.

7-10 Feedforward-plus-Feedback — Steam Flow plus Steam Pressure

A feedforward plus feedback arrangement is often used. One of the two most frequently used variations is shown in Figure 7-19. In this arrangement the steam flow (a) is the feedforward demand. The proportional multiplier function (b) is adjusted at the input of summer (c) so that a change in steam flow will produce the correct steady-state change in firing rate demand. The steam pressure controller (d) provides the correct adjustment of the firing rate demand for the necessary overfiring or underfiring to adjust energy storage.

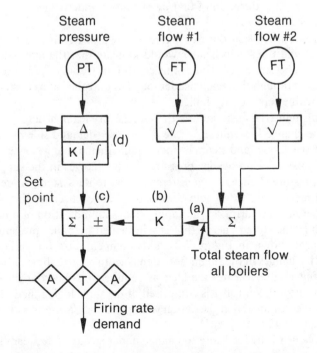

Figure 7-19 Feedforward-plus-Feedback Master Control

Using the previous example, the gain of the proportional multiplier (b) would be adjusted to (0.09/0.1) 0.9. The gain of the pressure controller would be (2.25/3.3) 0.68. In this type of system, most of the control action that would be supplied by the integral in a simple feedback control loop is supplied by the feedforward signal. Since the feedback portion should essentially produce no effect under steady-state conditions, it is necessary that the steam pressure controller be essentially proportional in nature.

In order that the steam pressure will eventually return to set point when the steam pressure deviation versus firing rate relationship is imperfect, a small amount of integral is needed. This should be an amount less than that indicated by the process time constant to avoid developing an unwanted integral signal during the steam pressure deviation. An integral setting of 0.05 to 0.1 repeats per minute is suggested.

Note that the steam flow signal is shown as the total steam flow for all boilers. Whenever more than one boiler feeds steam to a steam header, arranging one individual boiler feedforward from the steam flow of that boiler alone will result in unstable control due to positive feedback.

In the arrangement above, a change in the steam flow feedforward signal provides the necessary magnitude of the steady-state change in firing rate demand. An alternate feedforward application shown in Figure 7-20 exchanges the functions of the steam flow and steam pressure signals.

In this case the derivative input from steam flow into summer (c) is adjusted to provide the temporary overfiring or underfiring, with the steam pressure controller (d) adjusted to provide the necessary changes in the steady-state firing rate demand. The calibration of summer (c) does not include a bias, since under steady-state operation the output of summer (c) should equal the input from controller (d) with the steam flow derivative signal returning to zero.

Using the previous example, the steam pressure controller would have a gain of (9/3.3) 2.73. If the integral of the simple feedback controller were 0.25 repeats per minute, the integral would be 0.25 also for this arrangement. The derivative from steam flow would be adjusted so that the 2.25 percent magnitude under or overfiring, adjusted by proportional (b), and the necessary time duration of that overfiring or underfiring, adjusted by derivative (e), would be correct.

While the selection between the two alternates is user's choice, a high pressure drop between the boiler and the steam header would indicate that the alternate in Figure 7-20 would probably be a better choice. Under this condition the boiler pressure changes significantly with load even though the steam header may be controlled at a constant pressure. This is often the case with electric utility boilers.

In the normal industrial installation, the change in steam header pressure is almost entirely due to the change in steam flow rate on the user demand side of the header. If the pressure drop between boiler and steam header is high, then a change in steam flow rate may cause a larger change in steam header pressure due to changes in the supply side pressure drop. These two pressure changes mean different things to the system. A pressure change on the demand side means that firing rate should be changed because the user wants more steam. A change in pressure drop on the supply side is an indication of a needed change in firing rate to change the stored energy that is represented by boiler pressure.

The arrangement shown in Figure 7-21 adds compensation for variation in the heat content of the fuel. Changes in the steam flow signal act to change the steady-state firing rate demand. Overfiring or underfiring is added as required by the controller (d). The total proportional-plus-integral controller is comprised of the three logic functions marked (d). If fuel heating value is constant and at the design value, the analysis of this system is the same as that shown in Figure 7-19.

If the heating value of the fuel were to change under steady-state steam flow, the steam pressure would start to change due to a mismatch of heat input and output. The integral

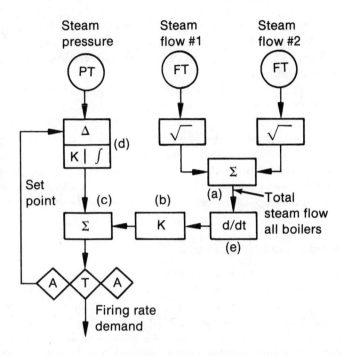

Figure 7-20 Alternate Feedforward-plus-Feedback Master Control

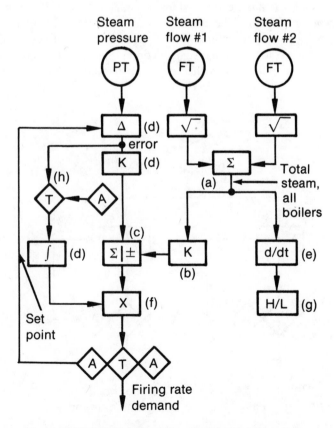

**Figure 7-21 Feedforward-plus-Feedback Master Control
(with automatic adaptive feedforward gain)**

function (d) would begin to generate a change in its output due to the steam pressure error from set point multiplied by the gain of the steam pressure controller. The output of the integral function (d) would then change the multiplication in the multiplier (f) to adjust for the change in the fuel heating value.

When the multiplication is correct so that the fuel heat input is in proper balance with the steam heat output, the steam pressure returns to set point, the steam pressure error becomes 0, and the action of the integral function (d) ceases to change the multiplication value.

In this application the integral tuning should be slow, with its output changing only on a sustained pressure deviation. The steam pressure deviations from normal steam flow changes should have no effect on the output of this integral function. The interlock functions (e), (g), and (h) are included to block integral action when the steam flow rate is changing. In effect the boiler is being used as a calorimeter to automatically keep the system in correct calibration.

7-11 Load Sharing of Multiple Boilers

The firing rate demand generated by the master controller raises and lowers the firing rate of all boilers connected to a steam header. The next control problem encountered is the proper sharing of the load between boilers. The simplest and most often used method is to leave the boiler load allocation to the judgment of the operator.

In practice a 0 to 100 percent firing rate demand signal is sent to a boiler master manual-automatic station. Incorporated into this station is a bias function. The bias function enables the boiler operator to add to or subtract from the master firing rate demand signal and thus alter this signal before it is transmitted to the individual boiler controls. Figure 7-22 shows the control functions that are involved for two boilers. In a similar manner this can be extended to three or more boilers. The operator should be instructed to maintain a 0 percent bias on one of the boiler master stations with input and output of equal value. On other stations the operator enters a desired bias signal in order that the output boiler firing rate will deviate from the firing rate demand by the amount of the plus or minus bias signal.

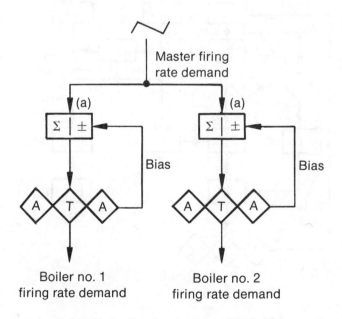

Figure 7-22 Manual Boiler Load Allocation

7-12 Automatic Compensation for the Number and Size of Boilers Participating

In the arrangement of Figure 7-22, an assumption is made in tuning the master steam pressure controller that the number and size of the boilers will always be the same. In many cases as the total load increases, additional boilers of indeterminate size may be put in operation. As the number and/or average size of the boilers connected change, the effect is to change the gain of the system. In this circumstance the optimum tuning of the master controller changes due to the change in total fuel flow capability. By modifying the load allocation functions as shown in Figure 7-23, automatic compensation for the number and average size of the boilers is obtained.

In this arrangement a summer function (b) and a high gain, fast integral controller function (c) is added. Assume that three boilers of the same capacity are feeding the steam header and that total steam flow and firing rate demand are at 60 percent. The feedback to the proportional-plus-integral control (c) from the summer (b) must also be at 60 percent, and the output of the controller would also be 60 percent boiler firing rate for the 60 percent steam flow. Under this condition, three 60 percent signals are inputs to the summer (b). Since the boilers are equal in size, a gain of 0.33 on each summer input will maintain the system in balance.

If one boiler is shut down with no change in total steam flow, one of the 60 percent inputs is removed from the summer (b). This immediately drops the summer output to 40 percent, causing an immediate increase in controller (c) output to 90 percent. Two 90 percent inputs to summer (b) with 0.33 input gain produces a summer output of 60 percent. The control loop is back in balance with a 60 percent input and 90 percent output from controller (c).

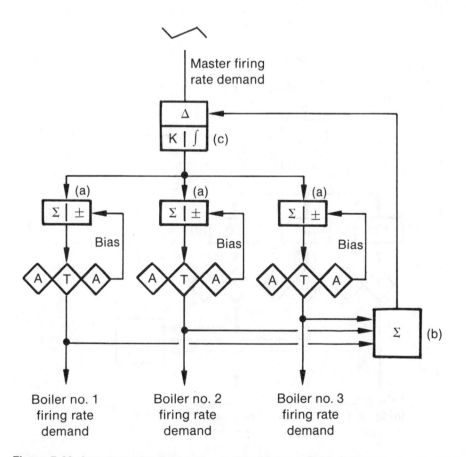

Figure 7-23 Automatic Compensation for Number and Size of Participating Boilers

Since there are now 2 boilers instead of three, each 1 percent change in master firing rate demand will result in 1.5 percent change in boiler firing rate demand. The arrangement produces correct results whether boilers are added or shut down.

Should one or more boilers be placed on manual control, its firing rate is constant with its input to summer (b) at a fixed value. Any change in master firing rate demand must then be balanced by an output change of summer (b) that does not include the input from the boiler on manual control. The relationship between the input and output changes of the controller (c) will then be the same as if the manually controlled boiler had been removed from service.

Should the boilers be of different sizes, gain values on the inputs of summer (b) are adjusted in accordance with boiler size. If three boilers are used and one is twice the capacity of the other two, then the summer gain used with the larger boiler would be 0.5 and the other two summer inputs would have gains of 0.25. Under the 60 percent load example, removing one of the smaller boilers from service would cause an immediate firing rate change on the two remaining boilers to 80 percent of full load. Whether the boilers are of the same or equal size, any adjustment of load allocation bias by the operator will cause an immediate readjustment to the firing rate of other operating boilers.

7-13 Preallocation of Boiler Load Based on Test Results

In some installations it may be desirable to remove the load allocation decision from the operator's function. The simplest method of automatically allocating load between the boilers is shown in Figure 7-24. In this control arrangement a function generator has been added for each boiler. The output signal of each function generator is some function of the input signal.

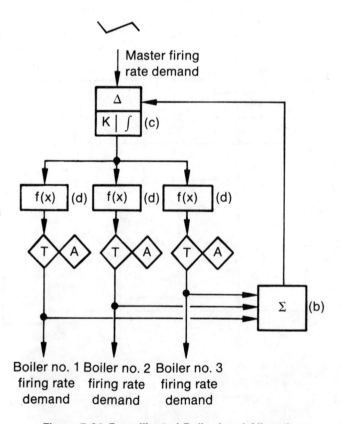

Figure 7-24 Precalibrated Boiler Load Allocation

In practice the boilers would be tested and the best overall load allocation would be determined. The individual boiler function generators would then be calibrated based on the desired function curves that were determined. The arrangement shown also has the capability to automatically compensate for the number of boilers in service. Should a boiler be removed from or added to service or placed in the manual control mode, the loading allocation of the boilers in operation should be approximately correct relative to the others in operation at the time.

7-14 Boiler Load Allocation on a Least-Cost Basis

A current state of the art technique for energy use efficiency is to allocate boiler loads on a least-cost basis. Such control is a part of the control systems that are generally called "energy management" control. The system is usually implemented with a digital computer, which sends boiler load allocation signals to the boiler master control. Figure 7-25 demonstrates the control logic of using the load allocation signals from the computer. The computer develops a bias signal to raise or lower the basic demand for boiler firing rate. A summer function (a) is used to combine the signals. The summer outputs are indicative of boiler load allocation on a least-cost or other "as-desired" basis. All summer input gain adjustments are 1.0.

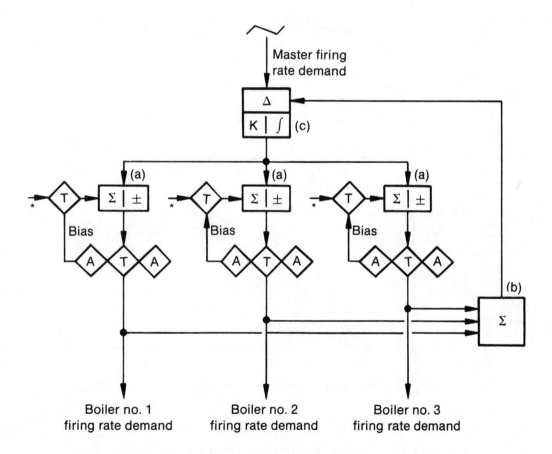

*From computerized load allocation system

Figure 7-25 Boiler Load Allocation on a Least-Cost Basis

The computations to develop the least-cost boiler load allocation signal include all applicable factors of the cost of boiler operation. These include the unit fuel cost, any special operation cost factors for a particular boiler, plus the incremental cost of input fuel energy relative to steam energy output. The boiler with the highest efficiency may not be the proper boiler for adding the next increment of total boiler load. The correct result is to load the boilers at equal incremental rates. The incremental heat rate is determined by the derivative of the input-output energy curve.

Figure 7-26 demonstrates the logic of this approach. Assume that there are two boilers and that test data produces curves of efficiency vs boiler load as shown. Boiler 1 shows an efficiency which at all loads is higher than that of boiler 2. From this it appears that it would always be more economical to load boiler 1 to 100 percent before loading boiler 2.

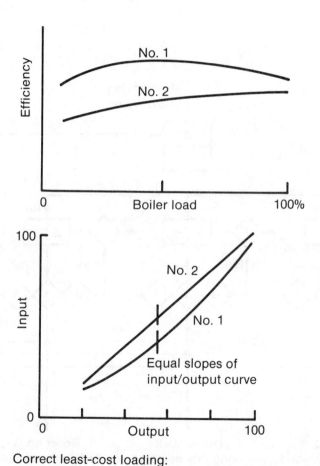

Correct least-cost loading:

Hold Boiler No. 2 at 20%; load No. 1 from 20% to 55%.
Hold Boiler No. 1 at 55%; load No. 2 from 20% to 100%.
Load Boiler No. 1 from 55% to 100%.

Figure 7-26 Least-Cost Boiler Loading

The input (fuel) vs output (steam) curve in Figure 7-26 shows clearly the error of this approach. For least-cost it is always necessary to obtain the next increment at lowest cost. In this case this would indicate loading boiler 1 from 20 percent to 55 percent, while the load of boiler 2 remains at 20 percent. If additional steam is required, the load of boiler 2 should then be increased up to 100 percent before any further changes are made to the load of boiler 1. Additional steam from this point is then obtained by loading boiler 1 from 55 percent to 100 percent. If there were a greater number of boilers, loading them all at equal incremental fuel rates would always require the lowest additional increment of energy input.

Section 8
Feedwater Supply and Boiler Water Circulation Systems

A typical boiler feedwater supply system consists of three basic parts. These are (1) a set of boiler feed pumps, (2) valves, feedwater piping, and headers, and (3) feedwater heaters. This system is supplied with condensate or chemically treated water at a relatively low temperature. The heaters and boiler feedwater pumps condition the water for proper admission to the boiler. The piping and headers connect the feedwater supply, heaters, valves, and pumps to the boiler.

8-1 The Basic System

Figure 8-1 shows a general arrangement of a basic feedwater supply system. The relatively cool water is admitted to the deaerating heater. Water leaving the heater is deposited into an integral heated-water storage tank. The storage tank is connected to the suction of the boiler feed pumps. At the discharge of the boiler feed pumps, a recirculation line containing a control or shutoff valve is connected. Downstream from this connection is a check valve between the pump and the feedwater supply header so that pressure from the supply header cannot return to the pump. With more than one pump connected, the check valve would be closed on any pump that is not in operation.

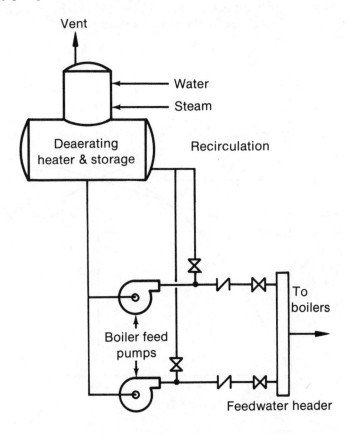

Figure 8-1 Boiler Feedwater Pumping and Heating System

The recirculation line is open at low flows and is sized for approximately 15 to 20 percent of pump capacity. The necessity for this recirculation is shown on Figure 8-2, which shows a typical set of characteristic curves for a constant speed boiler feed pump. Note that the power consumption is 60 to 70 percent of full load power at a 0 flow condition. With this condition of power input and no flow to dissipate the heat, the temperature would rise very rapidly and damage the pump. To avoid this potential damage, approximately 15 to 20 percent of the water is recirculated to keep the pump from overheating.

If the recirculation valve were operated manually, safe operation would dictate that the valve always be in the open position. Under such operation, only 80 to 85 percent of the pump capacity would be available for use. The recirculation valve can be automated with either a proportional or an on-off control. In either case, as the water flow demand is increased above the set point of the control, the recirculation valve is closed. The recirculation capacity is then available for useful work whenever the flow is above the set point range of the recirculation control.

8-2 Heating and Deaeration

If the boiler feedwater can be heated with steam that would otherwise be wasted, the increased feedwater temperature results in less fuel required to generate the steam. Another reason to heat the feedwater is the process called deaeration.

Before the water is put into the boiler, entrained or dissolved gases such as carbon dioxide and oxygen should be eliminated. If allowed to remain in the water, carbon dioxide will pass into the steam and turn into corrosive carbonic acid in the heat exchangers that use the steam. This corrodes the heat exchangers and the condensate return piping system. At the temperature of the boiler water, oxygen, if allowed to reach the boiler, can seriously corrode the boiler. Any remainder that leaves the boiler with the steam can corrode the heat exchangers and return lines.

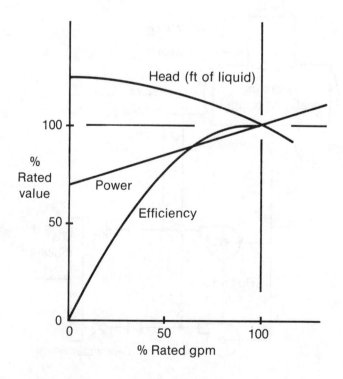

Figure 8-2 Characteristics of a Constant Speed Pump

Eliminating these entrained or dissolved gases before they can enter the boiler is called deaeration and is done by a deaerating heater. At colder temperatures, water can hold greater amounts of air, oxygen, carbon dioxide, or other gases in solution. These gases can be removed by vigorous boiling and a procedure for venting the gases to the atmosphere.

In the deaerating heater the water is heated by direct contact with steam by cascading the water-steam mixture over metal surfaces. In this way the water and steam become an intimate mixture at the boiling point. The entrained or dissolved gases are released from the water through this boiling and agitation. Because the pressure in the heater is above atmospheric pressure, the gases can then be vented to atmosphere.

Deaeration removes the oxygen and carbon dioxide as efficiently as mechanically possible. Scavenging chemicals are used in the boiler to eliminate the remaining traces.

8-3 The Boiler Feedwater Pump

The basic characteristic curves of a typical boiler feedwater pump, the power input curve, the head-capacity curve, and the efficiency curve, are shown in Figure 8-2. In all cases the "y" coordinate shows the percent of the rated head value. Rated load is shown as 100 percent on the "x" coordinate value. The units for the discharge pressure or "head" is in feet of the particular fluid being pumped. Since cold water has a higher density than hot water, the discharge pressure in psi will be higher when pumping colder water. The capacity on the "x" coordinate is a volumetric capacity, usually gpm (gallons per minute) for a boiler feed pump.

Boiler feed pumps may also be operated in a variable speed manner. In this case the speed would be varied with a variable speed motor, a magnetic or hydraulic coupling, or a steam turbine. If the pump is operated in a variable speed mode, the head-capacity curves are as shown in Figure 8-3. Slowing the pump speed also significantly improves part load efficiency and reduces pump power consumption.

In Figure 8-1 the boiler feed pump suction is shown connected to the tank of boiling water in the deaerating heater. To avoid the flashing of this boiling water into steam at the pump suction, there must be pressure at the pump suction in excess of the saturation pressure

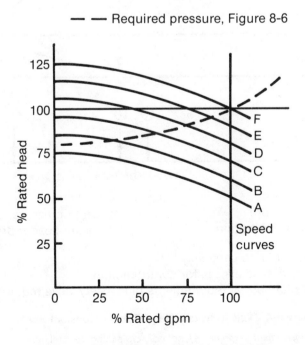

Figure 8-3 Speed Curves, Variable Speed Pump

of the boiling water. This is accomplished by locating the pump at a great enough difference in elevation below the heater so that there will always be this positive pressure difference. This is called NPSH (net positive suction head) and must also take into account the friction loss due to the flow from the heater to the pump.

8-4 The Flow Regulation System

Feedwater is continuously added to the boiler through piping to the steam drum as previously shown in Figure 2-14. At the point of entrance, it is necessary that the pressure in the feedwater system be slightly higher than the boiler drum pressure so that the water will flow to the drum. The feedwater regulating system controls the flow and dissipates the pressure difference between the boiler drum and the supply from the feedwater pump.

If the pump or pumps are driven by a constant-speed electric motor or turbine, the feedwater supply pressure, except for the piping friction losses, will follow the pump characteristic curves. If the steam header pressure is maintained at a constant value, the drum pressure will be at a higher pressure as determined by the friction losses through the super-heater (if the steam is superheated) and the valves and piping. In this case the feedwater flow is controlled by a standard heavy-duty control valve. The pressure drop dissipated by the valve is determined from a system head curve as shown in Figure 8-4.

While the actual flow is based on the pressure drops shown, other considerations are necessary in sizing the control valve. The lower pressure used in sizing is the pressure setting of the drum pressure relief valve, *since the system must be capable of adding water with the relief valve blowing.*

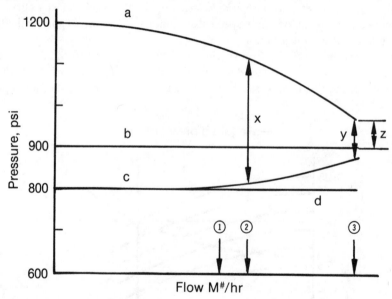

a. Feedwater pressure   c. Drum pressure
b. Drum safety valve setting   d. Steam header pressure

① Capacity at full firing capability
② Capacity above plus excess for control action
③ Capacity for all normal requirements plus safety valve requirements

Figure 8-4 Typical System Head Curve - Constant Speed

(Instrument Engineers' Handbook, Lipták and Venczel, eds.
© 1985 by Bela G. Lipták, Chilton Book Co., Radnor, PA)

Additional capacity is often designed into the control valve to accommodate the additional water that is lost from the system with the relief valve blowing. In some cases designing the control valve for the lower differential pressure and the added flow may result in an extremely oversized control valve. In this event it is better to size the control valve for good control and take care of the emergency conditions with manual or motor-operated bypass valves.

Some general guidelines for the control valve are as follows:

(1) Valve pressure drop should be not less than approximately 50 psi in order that the pressure drop will be a sufficient percentage of total system pressure drop to assure responsive control.

(2) Valve pressure standard should be based on the ASME boiler code if it is the first valve in the feedwater system away from the boiler drum. Otherwise the valve pressure standard can be based on the less stringent piping code.

(3) The valve body materials should be carefully selected, keeping in mind that very pure water is "metal hungry".

(4) As noted on the sytem head curve, the pressure drop varies with flow and tends to be considerably higher at low flows.

If the boiler feedwater pumps are variable-speed and more than one boiler is used with a feedwater pumping system, the feedwater pressure is normally controlled to a set point by changing the pump speed. In this case the system head curve is similar to that shown in Figure 8-5. With this arrangement pump power is saved by reducing the pump discharge pressure. In addition, the duty of the control valve is not as stringent since the valve pressure drop is lower and tends to be somewhat constant for all boiler loads. For design of the control valve, a minimum pressure drop of 50 psi is recommended.

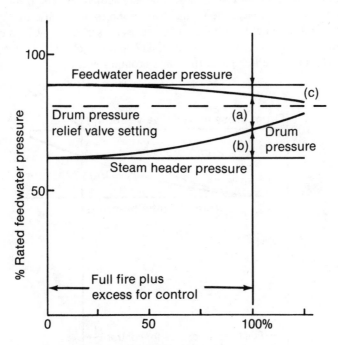

(a) Feedwater valve differential pressure
(b) Superheater differential pressure
(c) Feedwater piping differential pressure

Figure 8-5 Feedwater System Head Curve, Variable Speed Pumps

Utility boilers generally are unit-type operations with a single boiler, a single turbine, and a single set of boiler feed pumps. In this case the system head curve is as shown in Figure 8-6. The boiler feed pump speed is regulated to provide just enough pressure to force water into the pressurized boiler drum with no extra pressure for a control valve pressure drop. With this lower boiler feed pump discharge pressure, additional pumping power is saved. This depends to some extent, however, upon the method of steam temperature control. If spray water is used, then the system must be designed to provide the necessary pressure drop for the spray water control valve and the spray nozzle.

8-5 Shrink and Swell and Boiler Water Circulation

When the steam load on a boiler is increased, steam bubbles rise through the riser tubes of the boiler at a faster rate. The circulation of the water from the steam drum to the mud drum in the "downcomer" tubes and then as a mixture of steam and water up through the "riser" tubes has previously been discussed in Section 2. By the application of heat to the riser tubes, more steam bubbles are generated as more steam is demanded from the boiler. Similarly, reducing the heat applied to the riser tubes reduces steam bubble formation to satisfy the condition of reduced steam demand.

In the discussion of steam drum internal devices, it was demonstrated that there are two sections of space within the steam drum. One of these receives the steam-water mixture from the riser tubes, separates the water from the steam, and returns the remaining water to the water space. The feedwater is admitted into this relatively quiet water space. The boiler drum water level is measured also in this relatively quiet water space. From the level of the water in this space through the downcomers to the mud drum, there should be a very small number of rising steam bubbles.

If steam bubbles rise in boiler tubes, these tubes are acting as risers. If they are connected into the steam drum water space, the rising steam bubbles may cause the measured water level to appear unstable. Whether a boiler tube acts as a riser or a downcomer is dependent upon the amount of heat received by the tube. The amount of heat received is dependent upon the temperature of the flue gases that pass around the tube.

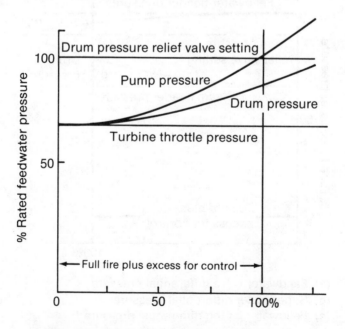

**Figure 8-6 Feedwater System Head Curve —
Single Boiler Variable Speed Pump, No Control Valve**

The flue gas baffle configuration affects the flue gas temperature in contact with specific tubes. Incorrect flue gas baffle design may cause hot flue gases to be applied to downcomer tubes, turning them incorrectly into risers and resulting in an unstable water level.

Figure 8-7 is based on an actual case of very poor gas baffle design and shows how the riser-downcomer identification can be confused by incorrect flue gas baffling. If this effect is major, as in this case, the flue gas baffling should probably be redesigned. If it is minor, it may be modified by a minor change in either the flue gas baffles or the steam drum internal baffles that separate the water space from the steam-water mixture space.

Assume that a boiler is being operated under steady-state conditions. At any point in time the boiler contains a certain mass of water and steam below the surface of the water in the steam drum. For this mass of water and steam there is an average mixture density. As long as the boiler steaming rate is constant, the steam-water mixture has the same volumetric proportions, and the average mixture density is constant.

Should the boiler load be increased, the concentration of steam bubbles under the water surface must increase. The result is that the volumetric proportions in the water-steam mixture changes and the average density of the mixture decreases. Since the mass of water and steam at this point has changed very insignificantly but the average density has decreased, the result is an immediate increase in the volume of the steam-water mixture. The only place the volume can expand is in the steam drum. This causes an immediate increase in the drum water level even though additional water has not been added. This effect of a sudden increase in drum water level as the steaming rate is increased is known as swell.

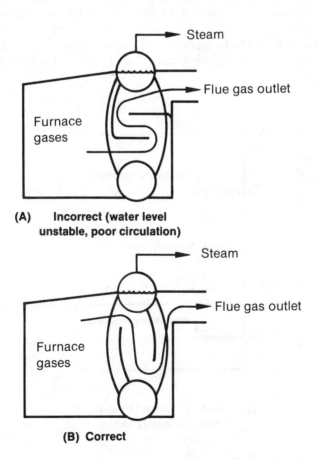

Figure 8-7 Flue Gas Baffles Affect Riser-Downcomer Identity

When the steam load is reduced, there are fewer steam bubbles in the mixture, the average density of the mixture increases and the volume of the steam-water mass decreases. The effect is an immediate reduction in the drum water level, although the mass of water and steam has not changed. This sudden reduction of drum water level on a decrease in steaming rate is called shrink.

The amount of water in the boiler at any given time is called the water inventory of the boiler. If the water has "swelled" due to an increase in steaming rate, the water inventory must be reduced to bring the drum water level down to the NWL (normal water level). If the water has "shrunk" due to a decrease in steaming rate, the water inventory must be increased to return the water level to the NWL.

Under steady steaming conditions there is thus less water in the boiler when steaming rate is high and more water in the boiler at a low steaming rate with the drum water level at the normal set point. This accounts for the fact that energy storage in the boiler water is higher at lower loads and lower at higher loads, as shown in Figure 7-11.

Figures 8-8 and 8-9 demonstrate the effect of changing load and inventory state. If steaming rate or load were increased and water flow to the boiler were immediately increased the same amount, the water inventory would remain constant. With this condition the drum level would be forced to remain in the swell condition. Only by delaying the water flow change can some of the excess inventory be converted to steam so that the drum water level can be returned to the set point. The reverse action that occurs as load is reduced requires an addition to inventory by delaying the reduction in water flow rate.

There are several factors that might change the apparent magnitude of the swell or shrink with a given load change. One of these is the size of the boiler steam drum as related to the water inventory and the change in steaming capacity. Because of greater drum volume, the swell or shrink will be less with a larger drum and no change in any of the other factors. With higher boiler pressure, steam density is greater, and the effect on mixture density, and thus swell and shrink, is less.

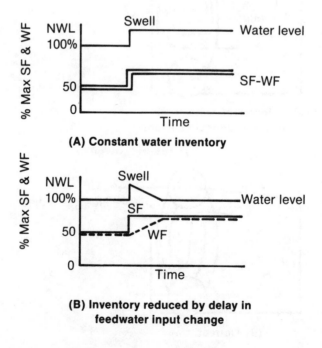

Figure 8-8 The Effect of Swell

The question is sometimes raised as to why the level in the water space changes since there should be no steam bubbles under the water level in this space. The answer is that the overall effect is felt in this space due to the increased water circulation rate. With increased flow through the downcomer tubes, the frictional pressure drop increases, causing a rise in the water level in the water space to balance the effect of changing mixture density in the steam-water mixture space.

If desired, the change in water inventory related to changes in steaming rate can be calculated from tests on the boiler or estimated based on the dimensions of the boiler drum. Referring again to Figures 8-8 and 8-9, the area of the triangle between the steam flow and water flow curves represents a particular mass of water. A boiler test can relate the time change and the change in steaming rate. The calculated amount should be approximately the same as that obtained using the observed swell and the drum dimensions.

This method is shown in Figure 8-10. From the pressure of the boiler, the boiler water density is determined. The observed swell and the drum dimensions can be used to calculate a

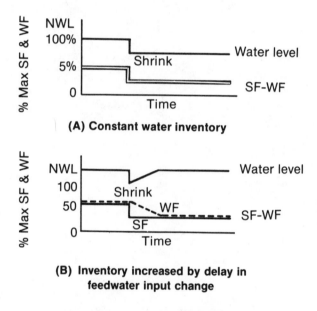

(A) Constant water inventory

(B) Inventory increased by delay in feedwater input change

Figure 8-9 The Effect of Shrink

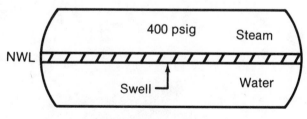

Assume steam drum, 20 ft long, 4 ft diam.
Swell = 2 in. (0.166 ft)
Volume of water (cu ft) = 20 × 4 × 0.166 = 13.28
Density @ 400 psig = 1/0.0194 = 51.55 lb/cu ft
Mass of water = 13.28 × 51.55 = 684.6 lbs

Figure 8-10 Change in Water Inventory

volume of water. This value multiplied by boiler water density should be approximately equal to the change in boiler water inventory for the change in boiler steaming rate. These values of the change in boiler water inventory related to changes in boiler firing rate are useful in analytical tuning procedures for feedwater control systems.

Section 9
Feedwater Chemical Balance and
Control of Boiler Blowdown

In all boilers a proper chemical balance must be maintained. The manner in which this is done can interact with the feedwater control system. Figure 9-1 demonstrates the chemical balance. The chemical control involves the chemical content of the boiler feedwater plus the much smaller quantity of boiler water conditioning chemicals.

The boiler steam scrubbers are intended to prevent any carryover of boiler water chemical content into the steam. All chemicals that enter the boiler through injection or in the feedwater must ultimately be removed in the boiler blowdown. Assuming a constant level of chemical concentration in the boiler water and also a constant concentration in the feedwater, the chemical concentration of the water in the boiler is determined by the ratio of blowdown flow to feedwater flow. If the average blowdown flow is 10 percent of feedwater flow, then the chemical concentration of the boiler water is 10 times that of the feedwater.

As the feedwater flow varies, the blowdown flow should also vary if the boiler water concentration ratio is to remain constant. In normal practice, the blowdown flow rate is adjusted periodically. During the interval between adjustments of blowdown flow rate, the chemical concentration in the boiler water may slowly rise or fall.

If the concentration in the boiler becomes too high, the boiler drum water level becomes unstable. The operating conditions of a boiler determine the maximum desired chemical concentration of the boiler water. Operation of the boiler at or near the maximum desired boiler water concentration results in minimum blowdown flow and reduced fuel loss. Assuming a saturated steam boiler, the relationship between percent blowdown and percent heat loss for different boiler pressures is shown in Figure 9-2. As the pressure of the boiler is increased, the blowdown heat loss increases due to the increased saturation temperature of the boiler water.

Two methods are used to remove the blowdown water from the boiler. A blowoff valve is connected to the lowest part of the mud drum. Periodic opening of this valve for a short time

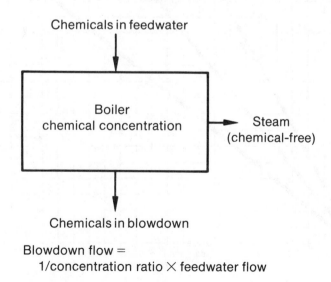

Blowdown flow =
1/concentration ratio × feedwater flow

Figure 9-1 Blowdown and Chemical Balance

period removes chemical sludge that has collected in the lowest part of the boiler. Blowdown water containing dissolved solids is removed through a continuous blowdown connection into the steam drum. This piping connection removes boiler water a short distance below the surface of the water in the steam drum.

Controlling the continuous blowdown is the normal method for controlling the chemical concentration. This can be done automatically using a boiler water conductivity measurement to control the blowdown flow rate and thus the chemical concentration. The automatic control method saves fuel by holding the average concentration closer to the maximum concentration and thus minimizing blowdown flow.

Since the limit is on maximum chemical concentration in the boiler water, lower chemical content in the feedwater allows a greater concentration ratio. The result is reduced blowdown and lower fuel loss due to blowdown. If the chemical content of the feedwater is high and this situation cannot be avoided through use of better water, then the blowdown will be relatively high. In some such cases it may be necessary to measure blowdown flow and add it to the steam output as part of the feedwater control strategy.

If the blowdown is not measured, its percentage of the feedwater flow can be estimated using chemical concentration ratios. A typical method uses the chemical concentration of chlorides in the feedwater and boiler water. Chlorides are selected since they do not change as a result of chemical reactions that may take place in the boiler water solution. Assume that the feedwater contains 20 ppm chlorides and the boiler water contains 600 ppm chlorides. The concentration ratio is 30 and the blowdown percentage is 1/30 or 3.33 percent. In order to obtain correct results with this method, the concentrations should be stable and not changing at the time the measurements are made.

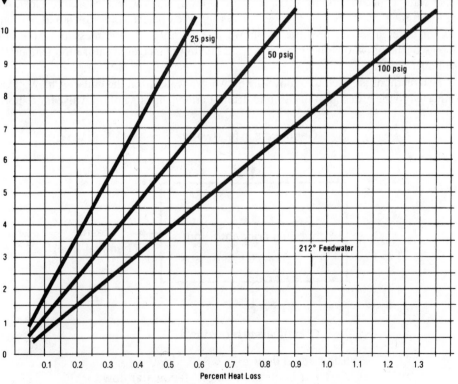

Figure 9-2 Effect of Boiler Pressure and % Blowdown on Blowdown Heat Loss

Section 10
Feedwater Control Systems

The flow of feedwater to the boiler drum is normally controlled in order to hold the level of the water in the steam drum as close as possible to the NWL (normal water level) set point. A typical level control loop measures level with a level sensor, processes this measurement in a proportional or proportional-plus-integral controller, and regulates the flow with a control valve. This typical control loop usually is inadequate for boiler drum level control.

This inadequacy results from the shrink and swell characteristics of the boiler, which produce level changes during boiler load changes in a direction opposite to that to which level would change under steady-state conditions. For this reason control from level alone will produce an incorrect control action anytime the boiler boiler load changes.

If, however, the boiler is small, has a relatively large water storage, and the load changes slowly, the simple single element level control may produce control performance that can be tolerated. With larger boilers, relatively less water storage, and faster load changes, the effect of shrink and swell tend to make the simpler systems inadequate.

10-1 Measurement and Indication of Boiler Drum Level

The basic indication of the drum water level is that shown in a sight gage glass connected to the boiler drum. The typical arrangement is shown in Figure 10-1. Since the configuration of the boiler and the distance of the boiler drum from the operator may not provide a useful "line-of-sight" indication, the gage glass image is usually projected with a periscope arrangement of mirrors so that the operator may easily view it. In some installations, the use of mirrors to project the water level image to a desired location for viewing may be mechanically complex or practically impossible, and other methods may be necessary. One such method is to use closed-circuit television; yet another is the use of a remote level indicator based on fiber optics.

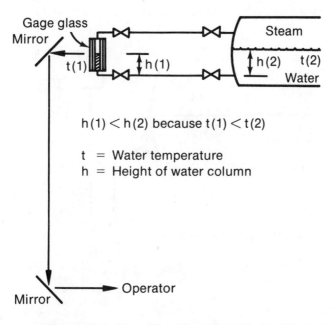

$h(1) < h(2)$ because $t(1) < t(2)$

t = Water temperature
h = Height of water column

Figure 10-1 Gage Glass Drum Level Indication

While the gage glass is the basic measurement, the indication it provides usually is in error to some degree and is not as correct as a properly calibrated level-measuring instrument. The basis for the error can be recognized from Figure 10-1. Condensate from cooling boiler steam circulates through the gage glass. This cooling of the steam and its condensate results in cooler water in the gage glass than in the boiler drum. The greater density of the cool water in the gage glass then shows a lower height water column to balance the column of water in the boiler drum.

Assuming a typical industrial boiler, the gage glass reading may be expected to read 1 to 3 inches of water below the actual level in the boiler drum. The deviation depends on the boiler pressure, the ambient temperature, plus piping and insulation between the boiler drum and the gage glass. For large high pressure electric utility boilers, the difference may be 5 to 7 inches. Some of the newer types of remote drum level-measuring instruments tend to compensate for the difference in readings described above.

When this fact is sufficiently understood, most of the error can be eliminated by physically lowering the gage glass. This fact must be well understood by anyone dealing with boiler drum level measurement and control, or much unproductive work will be performed trying to make a gage glass and measuring instrument agree.

A typical arrangement of a drum level measuring transmitter is shown in Figure 10-2. The transmitter is a differential pressure device in which the output signal increases as the differential pressure decreases. Typically the differential pressure range is approximately 30 inches with a zero suppression of several inches.

To determine the measuring instrument calibration, the necessary design data are the location of the upper and lower pressure taps into the boiler drum with respect to the normal water level, the operating pressure of the boiler drum, and the ambient temperature around the external piping. With these data and the desired range span of the tansmitter, the exact calibration can be calculated by using the standard thermodynamic properties of steam and water.

On the high-pressure side of the measuring device, the effective pressure equals boiler drum pressure plus the weight of a water column at ambient temperature, and having a length equal to the distance between the two drum pressure connections. On the low-pressure side, the effective pressure equals boiler drum pressure, plus the weight of a column of saturated steam having a length from the upper drum pressure connection to the water level,

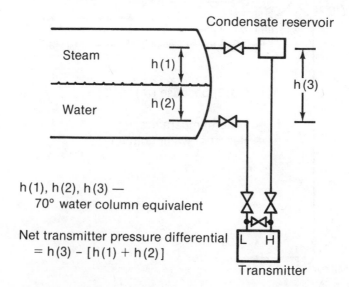

Figure 10-2 Drum Level Transmitter — Connection and Calibration

plus the weight of a column of water at saturation temperature having a length from the water level to the lower drum pressure connection.

Since the instrument measures differential pressure, the boiler drum pressure cancels out, leaving only the water column pressure difference. Since the density of saturated steam and water at saturation temperature changes as drum pressure changes, the level-calibration data will be correct at only a single boiler drum pressure. The signal from the measuring transmitter can be pressure compensated to be correct for all pressures by providing a drum pressure measurement and using it to multiply and bias the basic drum level measurement signal.

The bias is a calibration value in inches of water that represents the difference in the 100 percent (high) level calibration at a particular pressure compared to the calibration value at the base pressure condition. The multiplication changes the calibration span as the water in the drum changes density due to a change in pressure.

The values of bias and multiplication are based on a series of calculations that use different drum pressure data. Their effects are applied to the basic measurement signal to obtain a compensated signal. The bias is added or subtracted after the multiplying factor is applied to the basic signal.

Pressure compensation of the boiler drum level measurement is used almost universally on utility boiler applications. Such compensation is usually of little benefit in the normal industrial application where boilers generally operate at a constant pressure throughout the load range.

10-2 Feedwater Control Objectives

There are several basic objectives of feedwater control systems. Any judgment of performance of these systems should be related to how well these basic objectives are met. The feedwater control system must also cope with external influences or specific drum level characteristics that may tend to degrade the control performance. The major difficulties encountered are shrink and swell and variations in feedwater supply pressure.

In spite of these drawbacks, the system should meet the following objectives:

(1) Control the drum level to a set point.
(2) Minimize the interaction with the combustion control system.
(3) Make smooth changes in the boiler water inventory as boiler load changes.
(4) Properly balance the boiler steam output with the feedwater input.
(5) Compensate for feedwater pressure variation without process upset or set point shift.

Of particular importance is the elimination of interaction with the combustion control system. Such interaction is evidenced by uneven flow of feedwater. Such slugs of feedwater may cause upset to the steam pressure, thus resulting in firing rate changes with no changes in the steam flow rate. The firing rate changes cause shrink or swell and accentuates and continues the problem.

Just as there are basic objectives for good feedwater control, there is a basic pattern to the desired relationships of steam flow, feedwater flow, and boiler drum level that indicate good performance of a feedwater control system. The pattern of these relationships that indicates good feedwater control is shown in Figure 10-3. As steam flow increases, an increase in feedwater would be indicated if the boiler drum level did not swell. The drum level increase should cause a reduction in feedwater flow if the steam flow had not increased.

The proper adjustment of the feedwater control system balances these opposing influences so that the basic control objectives listed above are met. If the influence of drum level is too great, the initial control action will be to reduce feedwater flow. This will ultimately cause drum level to move beyond the control set point to make up for the lost water flow. If the influence of steam flow is too great, the initial control action will be to increase feedwater flow. This action will prolong the time period that the drum level is above set point. These two actions are shown in Figure 10-4.

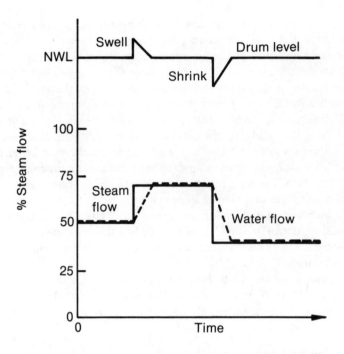

Figure 10-3 Desired Relationships to Meet Control Objectives

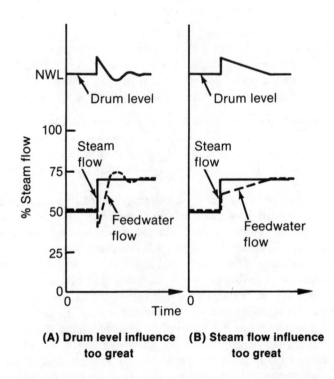

(A) Drum level influence   (B) Steam flow influence
**   too great   too great**

Figure 10-4 Proper Balance of Feedwater Control Influences

The correct action as shown in Figure 10-3 is for feedwater flow to make no immediate change, but rather to gradually change and reduce the water inventory as indicated by the return of the drum water level to set point.

10-3 Single-Element Feedwater Control

As indicated above, control of feedwater that relies only on a measurement of drum level will probably be adequate only on smaller boilers with a relatively large water volume and with relatively slow changes in load. The most elementary of such control is the "on-off" control normally used with a firetube boiler.

As shown in Figure 10-5, the level is held within about 3/4 inch. In the typical installation, feedwater flow is 0 or 100 percent depending on whether or not the feedwater pump is running. Theoretically, this would cause interaction with the firing rate control. Though the arrangement has this drawback, the water volume is large, shrink and swell account for only a small amount, steam flow changes are usually slow, and this type of feedwater control is usually adequate. This type of control certainly does not meet the control objectives or the desired pattern of the level and flows involved.

Smooth feeding of the feedwater would tend to eliminate the interactions with combustion control and improve boiler efficiency. More power would be needed for the feedwater pump since it would operate continuously. Even though the potential gain may be small, complete operating data on any particular installation can help determine whether control improvement would be economically feasible. If an economizer is used, the feedwater control must be a continuous flow-modulating type to insure that the feedwater flow continues through the economizer whenever the boiler is being fired. Continuous water flow is necessary to avoid damage to the economizer.

There are two general types of mechanical feedwater regulators that operate on the single element (drum level only) basis. These are proportional controls with a permanent level set point offset that is associated with each feedwater control valve position.

The first of these, called a thermostatic type, is an inclined tube, the ends of which are connected to the steam and water space in the boiler drum. Condensing steam in the upper

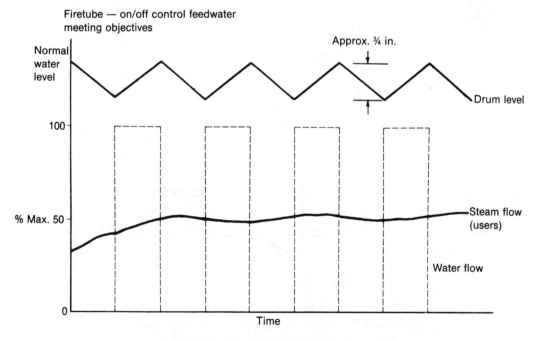

Figure 10-5 On-Off Feedwater Control Action

part of the tube causes circulation and a corresponding water level to be established inside the tube. As drum water level falls, there is more steam in the tube which causes it to heat and expand in length. The feedwater control valve is mechanically linked to one end of the tube, causing the valve to open or close as the tube expands or contracts. The elevation of the tube must be carefully located to obtain the proper set point range. The controller gain can be changed only by altering the slope of the inclined tube.

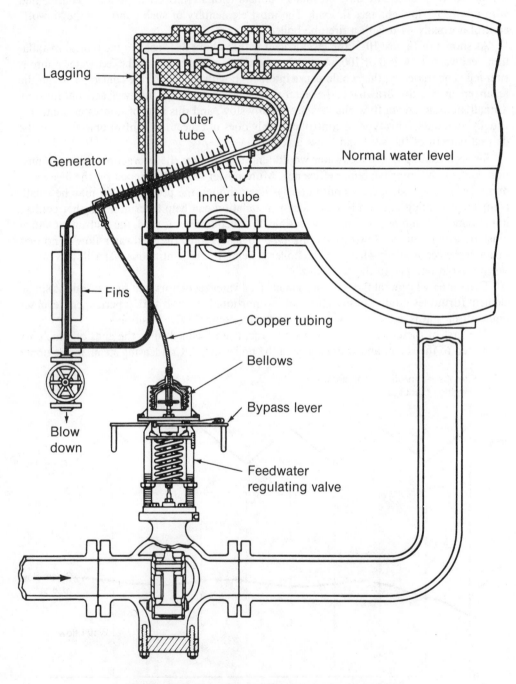

Figure 10-6 Thermohydraulic Feedwater Regulator

(From *Steam, Its Generation and Use*, ©Babcock and Wilcox)

The second of the mechanical feedwater regulators is known as the thermohydraulic type. A cross section of this type of regulator is shown in Figure 10-6. As in the thermostatic type, this regulator also has an inclined tube connected to the steam and water space of the boiler. Around this tube is a closed jacket that contains water or other vaporizing fluid. This outer closed jacket is connected by copper tubing to a bellows-operated control valve. As the drum level falls, the water in the outer jacket is exposed to greater heat transfer, the outer jacket pressure increases, and the bellows-operated control valve opens. The gain of this control is fixed by the slope of the generating tube. The normal shift of drum level set point for these two proportional regulators is approximately 4 inches from low load to high load.

The performance of these two devices is very similar and can be recognized from a chart of their typical results shown in Figure 10-7.

While there are significant inadequacies in these results as compared to the basic objectives of feedwater control they can be tolerated in most installations of smaller size boilers with slower load changes. The most serious drawback is the interaction with the firing rate control since this would tend to degrade the boiler efficiency.

An improvement in performance over that of the single-element mechanical system can usually be obtained by using a standard feedback control loop as shown in Figure 10-8. Being able to reduce the controller gain, results in less interaction. If the controller were proportional-only as in the mechanical regulator, a greater drum level offset would occur as boiler load changed

To avoid this unsatisfactory condition , integral control is added. The integral effect must be quite slow since the level signal moves incorrectly at times of boiler drum level swell and shrink. The result as shown in Figure 10-9 is a compromise that is an improvement over the mechanical control but has less than the desired performance. The specific improvement is that the incorrect action of feedwater flow during load changes is reduced, and the system is generally more stable and less interactive with the firing rate control.

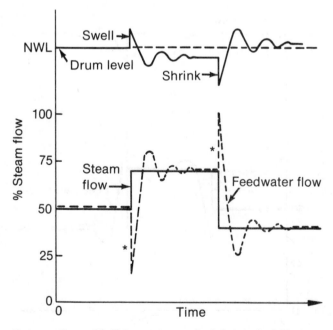

*Interaction with firing rate control due to imbalance between steam flow and feedwater flow.

Figure 10-7 Control Action of Mechanical Single-Element Feedwater Regulator

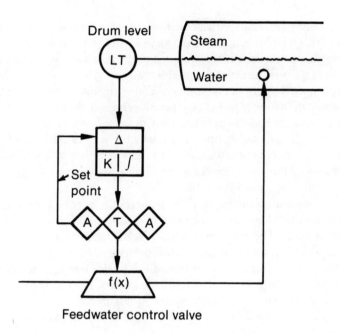

Figure 10-8 Simple Feedback Feedwater Control (Single Element)

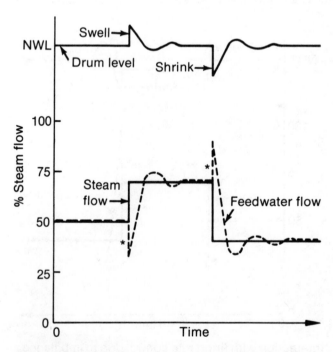

*Interaction with firing rate control due to imbalance
between steam flow and feedwater flow.

Figure 10-9 Control Action of Simple Feedback Feedwater Control (Single Element)

The controlled device, whether a control valve or pump speed control device, should have a linear signal vs flow characteristic as shown in Figure 10-10. The basic reason for this is that drum level deviations around the set point represent a specific quantity of water over the entire boiler load range. In the case of signal vs pump speed, this must be nonlinear as shown in order that the control signal vs water flow is as linear as possible. The large increase in pump speed for an initial small increase in control signal brings the pump speed up to the required pressure for admitting water into the boiler.

10-4 Two-Element Feedwater Control

A two-element feedwater control system is shown in Figure 10-11. This is easily recognized as a standard feedforward-plus-feedback control loop. In this case steam flow is the feedforward signal that anticipates a need for additional feedwater flow. The feedback control from drum water level is shown as proportional-only control. The control valve is

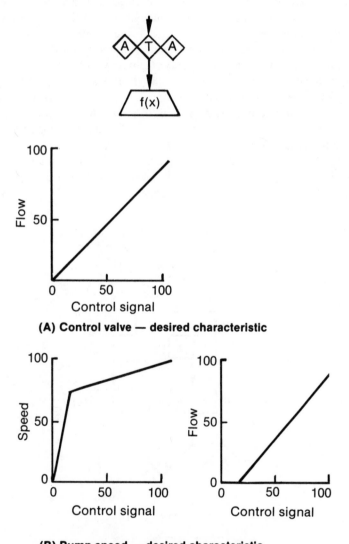

(A) Control valve — desired characteristic

(B) Pump speed — desired characteristic

Figure 10-10 Feedwater Control Device Flow Characteristics

characterized so that the control signal vs feedwater flow is linear. For this system to perform properly and hold the drum level at the set point, it is necessary that the differential pressure across the feedwater control valve be predictable at each flow and that the control valve signal vs flow relationship does not change.

Under the above conditions the system can be tuned so that performance such as shown in Figure 10-12 can be achieved. This performance is recognized as having the desired pattern of flow and level relationships, and such performance meets the boiler feedwater control objectives that have been stated previously.

Tuning such a system for proper action during the shrink and swell period requires the correct balance between the effects of steam flow and drum level. As stated before, the desired condition is for water flow to hold its flow rate during a load change and change only as the drum level begins to return to its set point. In this manner water inventory is smoothly adjusted to its new desired value.

Since the drum level control signal calls for a feedwater decrease as the steam flow signal is calling for an increase, the proper gain setting on steam flow and drum level should cause them to offset each other and affect no immediate change in the water flow control valve signal. As the drum level begins to change, the feedwater valve control signal is changed to keep the system in continuous balance until steam flow and water flow are again equal and drum level is at the set point. At this point, since steam flow and water flow are equal, there is no driving force to cause further changes in boiler drum water level.

Assume that the steam flow signal range is 0 to 100 percent for 0 to 200,000 lbs/hr. Assume that the feedwater control valve is sized for a maximum flow of 250,000 lbs/hr and that the control signal of 0 to 100 percent is linear with respect to this 250,000 lbs/hr flow. The correct feedforward gain in item (a) is 0.8. In this way, if steam flow were at 200,000 lbs/hr, the 100 percent signal would be multiplied by 0.8 before going to the control valve. The resulting 80 percent of the 250,000 lbs/hr control valve capacity would provide the correct 200,000 lbs/hr of feedwater to match the steam flow.

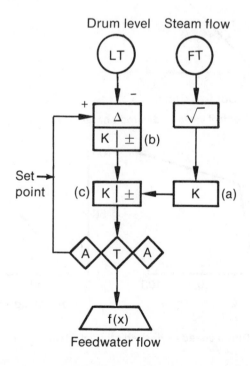

Figure 10-11 Two-Element Feedwater Control System (Feedforward-plus-Feedback)

Assume also that the range of the drum level transmitter is 30 inches with 0 or normal water level at the center with a 50 percent signal. The 0 percent signal corresponds to minus 15 inches and the 100 percent signal to plus 15 inches. A test of the particular boiler indicates that the boiler swell is 2.5 inches as the steam flow is rapidly increased by 20,000 lbs/hr.

In determining the proper gains so that the steam flow and boiler drum level signals may exactly offset each other, the steam flow gain of 0.8 must remain fixed to correctly match the steam flow range to the control valve range (Figure 10-11, item (a)). The drum level gain in item (b) must therefore be adjusted to match the effect of the 0.8 steam flow gain. Since the steam flow change is 20,000 lbs/hr or 10 percent of maximum, the net effect from steam flow is 0.8×10 percent or 8 percent. The drum level has changed 2.5 inches, which is 8.33 percent (2.5/30) of its range. To match the steam flow effect, the gain on the drum level signal must be 0.96 (8/8.33). This gain is then applied to the transmitted signal in item (b) in Figure 10-11. If the swell had been 1.5 inches, the drum level gain would be 1.6 (0.08/0.05).

The calibration of the system includes a bias adjustment to the output signal of item (b) and item (c). The effect of the output of item (b) should have both positive and negative possibility. If the signals in a particular system can have only positive values, the effective output of item (b) should be 50 percent so that it can change in both upward and downward direction. The 50 percent positive bias of the output signal from item (b) adjusts the normal water level 0 percent output signal to 50 percent. The 50 percent signal combines with the steam flow signal from item (a). This requires a negative 50 percent bias to the output signal of item (c) so that the control valve signal will be correct. These two bias values would be 0 for systems that can work with both positive and negative values.

While the system shown will achieve all of the desired control objectives under the conditions specified, it has a serious drawback if the feedwater control valve pressure differential and thus the control valve flow characteristic is not always the same. Figure 10-12B demonstrates how the performance is seriously degraded by variations in feedwater pressure. Such feedwater pressure variations change the relationship between steam flow and feedwater flow. Boiler drum level is then forced to develop an offset from set point in order to bring the steam flow and feedwater flow into balance. Under conditions of unpredictable or variable feedwater pressure, three-element feedwater control is necessary if the desired results are to be achieved.

10-5 Three-Element Feedwater Control

Two-element feedwater control uses the two measurements of steam flow and boiler drum level. Three-element control adds the measurement of feedwater flow into the control strategy. In the preceding paragraphs, it was demonstrated that an unpredictable control signal vs feedwater flow characteristic seriously degraded control performance. Three-element control assures that the signal vs feedwater flow will have a constant relationship by replacing the open-loop flow characteristic of the feedwater control valve with a closed-loop feedback control of feedwater flow.

There is more than one way to arrange a three-element feedwater control system, but the most common can be described as a feedforward-plus-feedback cascade control. This arrangement is shown in Figure 10-13. Two other arrangements are shown in Figures 10-14 and 10-15.

The most common arrangement (shown in Figure 10-13) is tuned using the same thought process that is used with the two-element system except that the feedwater flow control is substituted for the control valve characteristic. If the feedwater flow measurement is 0 to 250,000 lbs/hr, and the steam flow measurement is 0 to 200,000 lbs/hr, the gain of item (a) would be 0.8. With the same drum level transmitter and swell effect of 8.33 percent of drum level span for a 10 percent change in steam flow rate, the gain of the proportional level control (item b) would be 0.96.

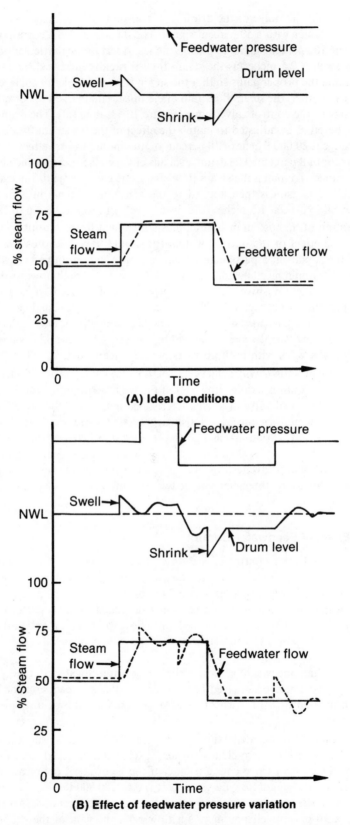

Figure 10-12 Performance of Two-Element Feedwater Control

The tuning of the feedwater control loop would be typical of a flow control loop. A reasonable starting point for the tuning of the flow controller (item c) would be a gain of 0.5 and an integral setting of 4 repeats per minute. The tuning and calibration constants of the control arrangements in Figures 10-14 and 10-15 should be identical to those of the arrangement in Figure 10-13.

Properly tuned, the performance of the three-element control should appear approximately as shown in Figure 10-16. Note that this performance, although the feedwater supply pressure may vary, meets all the control objectives and contains the correct pattern of relationships between steam flow, feedwater flow and boiler drum level.

In some applications, the drum level measurement contains process noise from slight to severe fluctuations in drum level. In addition, the measured steam flow may contain fluctuations or process noise. These noise components may cancel or add depending on their characteristics. If they add, they often create control stability problems. The control arrangement in Figure 10-14 offers more flexibility in eliminating the undesirable process noise effects. In this arrangement the adjustable time function (f/t) filters the feedback signal to the controller and thus removes the major share of the problem.

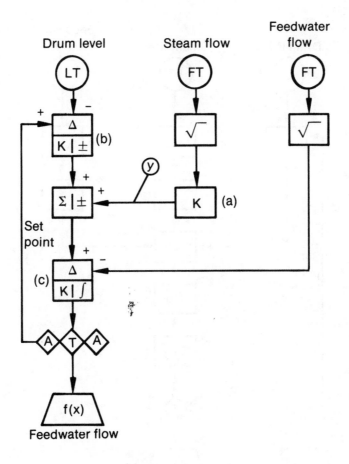

**Figure 10-13 Three-Element Feedwater Control
(Feedforward, Feedback, plus Cascade)**

10-6 Control Refinements and Special Control Problems

In Figure 10-16 a range of feedwater pressure is shown. Each change in feedwater pressure causes a change in the control valve differential pressure, effectively changing the control valve maximum capacity. The result is a change to the optimum tuning of the feedwater flow control loop. From a practical standpoint, the control is often detuned to some extent to allow for such variations.

Should the differential pressure, and thus the effective control valve capacity, change an extreme amount, it may be necessary to adaptively change the tuning gain of the flow control loop. Figure 10-17 shows a control arrangement that adds a fourth element of feedwater pressure. Such an arrangement is very seldom required, but is available to solve such problems or to eliminate the need for compromise tuning.

In some cases the blowdown flow is variable and of a relatively high percentage that significantly changes the relationship between steam flow and feedwater flow. In this case it may be necessary or desirable to add blowdown flow as a fourth element to steam flow, as shown in Figure 10-18.

On installations where unmeasured saturated steam is taken directly from the boiler drum for soot blowing, an imbalance between measured steam and feedwater flow may cause drum level offsets from set point. A compromise solution to this problem is to add integral

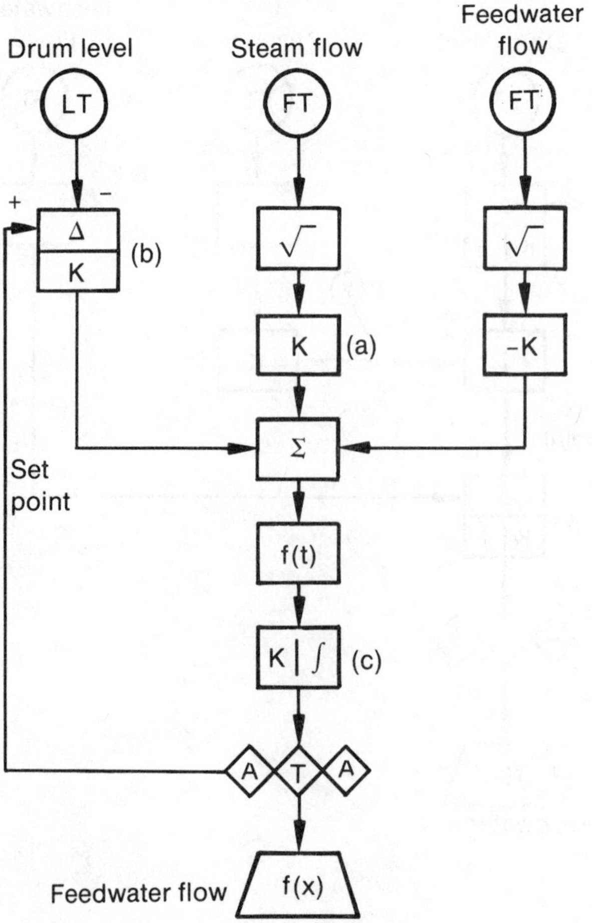

Figure 10-14 Three-Element Feedwater Control (Alternate)

control to the level controller and adjust the integral setting so that there is very little integral effect. An integral setting of 0.05 to 0.1 repeats per minute is suggested. Theoretically because of the "wrong way" action due to swell and shrink any integral effect adds instability into the system.

A fine point to the proper balance between steam flow and water flow for boilers with very rapid load changes is based on the changes in boiler pressure. The steam flow measurement concerns only the steam flowing through the steam line without consideration of the steam flow being added to or subtracted from the energy storage. In fact, the steam actually being generated includes the steam being generated that contributes to raising or reducing boiler pressure. Such steam is indicated by a derivative (d/dt) of the drum pressure. A control arrangement that includes this steam is shown in Figure 10-19.

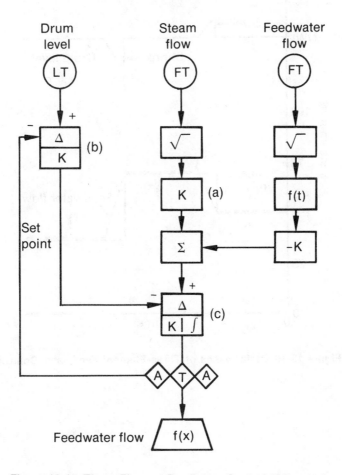

Figure 10-15 Three-Element Feedwater Control (Alternate)

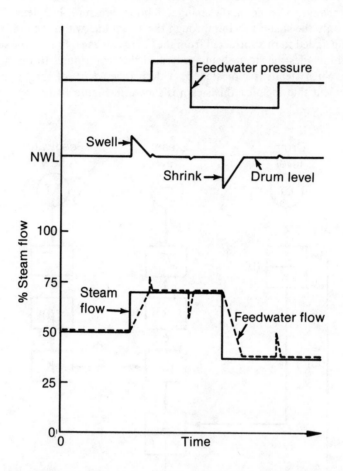

Figure 10-16 Performance of Three-Element Feedwater Control

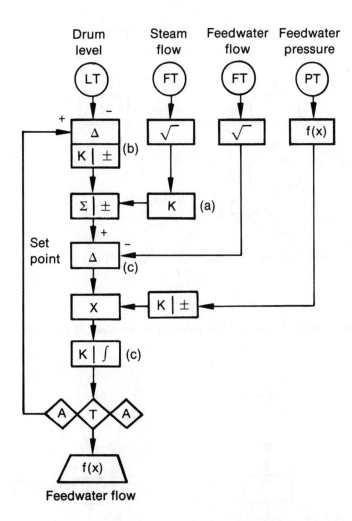

**Figure 10-17 Three-Element Feedwater Control
(Flow Controller Gain Pressure Adaptive)**

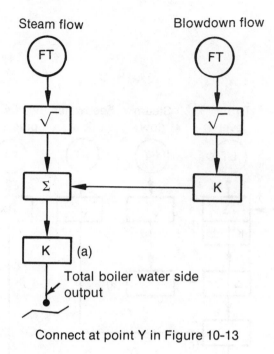

Figure 10-18 Alternate to Steam Flow Input for Three-Element Feedwater Control

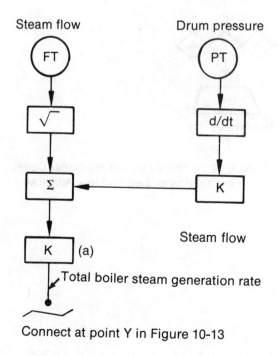

Figure 10-19 Alternate for Steam Flow Input for Three-Element Feedwater Control

Section 11
Boiler Draft Systems

The term "draft" has many different meanings but in this context it is defined as a "current of air". Associated with this meaning is its definition as an air or flue gas pressure that is slightly negative with respect to atmospheric pressure. When in a conduit connected to atmospheric pressure, such negative pressure would create a current of air or flue gas. The common definition of draft loss is the difference in draft or pressure resulting from the flow of the air or flue gas. These terms are used in connection with pressure and flow measurements of boiler flue gas and combustion air.

11-1 Draft Losses in Boilers

Boiler combustion air and flue gas flow through a system that includes the boiler furnace, ductwork, and various types of heat transfer surface. The driving force for this flow is an air or flue gas pressure or draft. The combustion air originates in the atmosphere and eventually is exhausted to the atmosphere. The total draft or pressure is divided up by all those elements in the flow path that tend to resist or obstruct the flow. The amount of the pressure differential for each of these elements is called its draft loss.

The draft loss for these flow restrictors generally follows the same differential pressure vs flow/square root relationship as an orifice in a piping system. The relationship may deviate to some extent from a true square root due to variations of specific volume over the flow range in addition to innate characteristics of the restriction. Figure 11-1 demonstrates the total and divided draft losses with respect to flow for a simple boiler without heat recovery equipment. Adding heat recovery equipment such as an air preheater requires additional draft loss on both the combustion air side and the flue gas side of the heat exchanger. This addition is shown in Figure 11-2.

Other factors may cause an apparent deviation in draft loss that is entirely unrelated to flow. When air and flue gases are heated, they become lighter and tend to rise. In boilers this

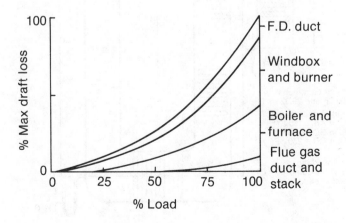

Figure 11-1. Draft Losses for All Boilers (No Air Preheater)

is known as the stack effect. If the flow is moving up, this produces an apparent draft loss higher than that accounted for by flow alone. If the flow is moving down, the effect is a reduction in the measured draft loss with respect to that produced by the flow alone. The amount of this stack effect in each case is affected only by the temperature (and thus specific volume) of the gases. Figure 11-3 demonstrates this stack effect in boilers.

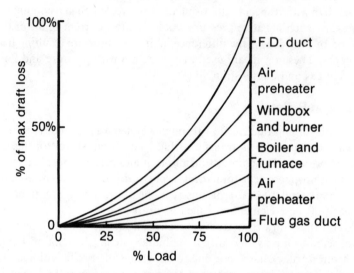

Figure 11-2. Draft Losses for All Boilers (Including Air Preheater)

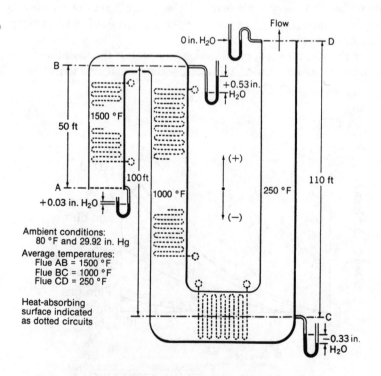

Figure 11-3 Stack Effect in Boilers

(From *Steam, Its Generation and Use*, ©Babcock and Wilcox)

11-2 Natural Draft and Forced Draft

Natural draft is a term used when the air flow through the boiler is a result of the stack effect in a chimney or stack. The stack is connected to the flue gas passage of the boiler. If the flue gas specific volume (primarily due to its temperature) is less than that of the outside atmosphere, the flue gases at the top of the stack will rise, creating a suction that will induce combustion air flow through the boiler. The draft that will be produced by a stack is a function of the height of the stack and the flue gas temperature. This relationship is shown in Figure 11-4.

Theoretically, the draft produced is independent of the stack diameter. In practical terms, however, the stack diameter must be sized so that the draft produced by the stack is not appreciably used by friction from the flow of the flue gases up the stack.

From the above it follows that some stack draft will always be present and can be used alone or in combination with a combustion air fan or fans. Generally, natural draft alone produces much less draft and is available for much lower draft losses than those available with mechanical draft. The result is poorer heat transfer and lowered boiler efficiency. On an economic basis, natural draft should be used to supplement to rather than as a substitute for mechanical draft.

Mechanical draft is that produced by combustion air fans. In the case of boilers, a fan or air blower that takes suction from the atmosphere and forces combustion air through the system is called a forced draft fan. A fan at the end of the boiler flow system path that takes its suction from the boiler flue gas stream and discharges the flue gas to the stack is called an induced draft fan. The static pressure and flow characteristics of fans result from the specific design of the particular fan. The fan combinations available to the boiler system designer are (a) forced draft plus stack, (b) forced draft and induced draft plus stack, and (c) induced draft plus stack.

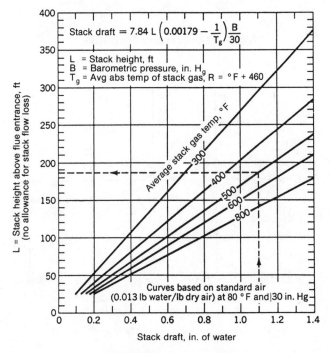

Stack height required for a range of stack drafts and average stack gas temperatures.

Figure 11-4 Stack Draft

(From *Steam, Its Generation and Use*, ©Babcock and Wilcox)

11-3 Pressure-Fired Boilers

A boiler system that contains no induced draft fan and whose furnace may operate under a positive pressure over some portion of the load range is called a pressure-fired boiler. In this type of boiler, the pressure in the furnace varies as the load is changed. This is due to the variation in the various draft losses with respect to boiler load.

Figure 11-5 represents the physical arrangement of such a boiler system. A key point with such boilers is that the furnace must be airtight or flue gastight. This is necessary so that the very hot flue gas of the furnace can not leak to the atmosphere. A small leak under such circumstances will deteriorate the material around it, eventually destroying the furnace walls and creating an operational hazard. Such furnaces are made pressure-tight with a welded inner casing or a welded seal between the furnace wall steam generating tubes.

A profile of the draft and pressure of a simple pressure-fired boiler without heat recovery equipment is shown in Figure 11-6. Note that at 70 percent boiler load the draft losses are approximately 50 percent of the full load draft losses. The draft or negative pressure at the

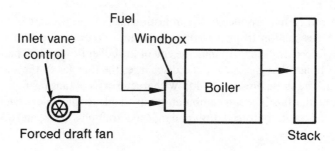

Figure 11-5 Pressure-Fired Boiler

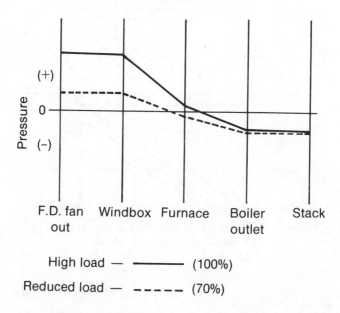

Figure 11-6 Profile of Pressure and Draft of a Pressure-Fired Boiler
(Typical — No Air Preheater)

right side of the profile results from the natural draft of the stack. If the stack draft were less, the entire profile would be raised. In this case the result of the natural draft is that the furnace is under pressure only at the higher loads.

A pressure and draft profile for a boiler with an air preheater type of heat recovery equipment is shown in Figure 11-7. The result of this is that the additional draft losses cause the furnace to be operated under pressure except at the very low boiler loads. If an economizer had been used for heat recovery instead of an air preheater, then there would be no additional draft loss on the combustion air side flow path.

11-4 Balanced Draft Boilers

In many cases furnaces cannot be operated under pressure because of leakage around the fuel burning equipment. An example of this is stoker firing of solid fuel. In other cases systems have been designed for negative furnace pressure operation to reduce furnace maintenance, or they were designed and constructed before pressure furnace technology was developed. Such boiler systems usually rely on the use of an induced draft fan in combination with a forced draft fan. In these cases the induced draft fan is used to reduce the furnace pressure and assure that it is always negative with respect to atmospheric pressure.

Such systems are called balanced draft sytems and have an arrangement as shown in Figure 11-8. In the balanced draft system, operating the furnace under a negative pressure assures that any leakage will be relatively cool combustion air leaking into the furnace instead of very hot combustion gases leaking out. In normal practice the furnace pressure or draft is controlled to a very slightly negative pressure set point by regulating either or both the forced and the induced draft fans. In this way any atmospheric air leaking into the furnace or boiler is minimized.

In the balanced draft boiler the forced and the induced draft fans share the load of moving the combustion air and flue gases through the system. The balance point is the pressure or draft in the furnace. This pressure level is determined by the relative amounts of "push" and "pull" of the forced and induced drafts, respectively.

A pressure and draft profile of a balanced draft boiler is shown in Figure 11-9. Note that the furnace draft is slightly negative for all boiler loads. Adding a heat recovery combustion

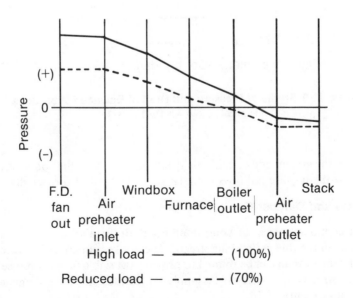

Figure 11-7 Profile of Pressure and Draft of a Pressure-Fired Boiler (Typical — Includes Air Preheater)

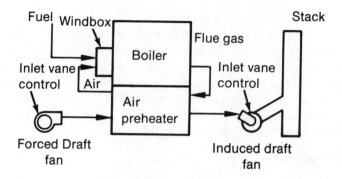

Figure 11-8 Balanced Draft Boiler (With Air Preheater)

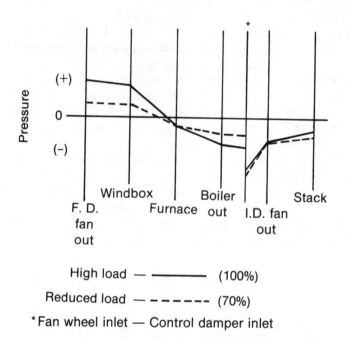

High load — ———— (100%)

Reduced load — – – – – – – (70%)

*Fan wheel inlet — Control damper inlet

**Figure 11-9 Profile of Pressure and Draft of Balanced Draft Boiler
(Typical — No Air Preheater)**

air preheater to the system adds additional draft losses to both the combustion air and the flue gas sides of the furnace but does not change the controlled furnace draft set point.

11-5 Dampers and Damper Control Devices

The most common device for controlling boiler drafts and air flows is some form of damper or vane in the flue gas and air stream. These take numerous forms ranging from a single-bladed damper than can be rotated to provide a variable flow resistance to a complex multibladed control vane or louver. A simple representation of the damper as a control device is shown in Figure 11-10.

Generally, all dampers have a nonlinear flow vs damper opening characteristic as shown. A multibladed damper tends to be more linear than a single-bladed damper. Dampers with

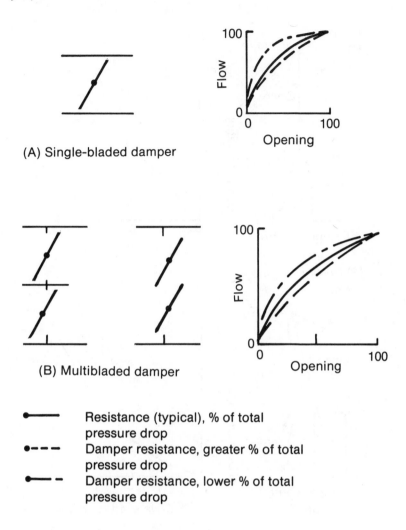

(A) Single-bladed damper

(B) Multibladed damper

Resistance (typical), % of total pressure drop

Damper resistance, greater % of total pressure drop

Damper resistance, lower % of total pressure drop

Figure 11-10 Damper Characteristic Flow (Typical)

adjacent blades rotating in a counter direction tend to be more linear that when all blades rotate in the same direction. In both cases the actual flow characteristic is more or less linear depending upon the ratio of the pressure drop across the damper to the total system pressure drop. The flow vs opening characteristic is more linear with more of the system pressure drop occurring at the damper and less linear with the damper accounting for less of the total system pressure drop.

For air flow or draft control it is very desirable to have linear characteristics of flue gas or air flow vs the control signal. A method of using a linear damper actuator while effectively altering the basic damper flow characteristic is called linkage angularity. This term applies to the modification of the linkage angles and lengths between the damper and its actuator.

The use of *linkage angularity* is demonstrated in Figure 11-11. In (A) the linkage driving and driven arms are of equal length and parallel. The angular rotation of the actuator shaft and the damper shaft will be equal and thus linear with respect to each other. By altering the lengths of the driven and driving arms and changing their relative angles as shown in (B), the actuator motion will be nonlinear with respect to the damper. The result is a nonlinearity, which, when combined with the opposite nonlinearity of the damper flow characteristic, tends to produce a linear actuator motion vs flow.

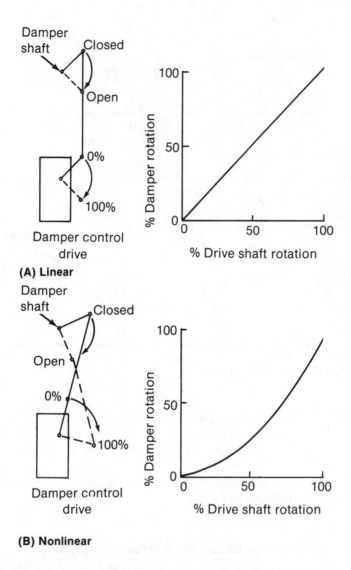

Figure 11-11 Damper Linkage Angularity

If this mechanism is to be used, the starting angles of the linkage must be fully adjustable in small increments. For this reason, damper actuators are often furnished with a spline on the output shaft and drive arm. It should also be noted that linkage angularity applies a nonlinear torque to the damper shaft and may be used to amplify the torque over a portion of the range while reducing it over another portion of the range.

The flow vs control signal can also be made linear by making the control signal versus actuator position nonlinear. In this way the basic nonlinear damper flow characteristic can be combined with an opposite nonlinear actuator characteristic to provide a linear flow vs control signal characteristic. This linearization method uses an actuator positioner that incorporates a nonlinear cam. The shape of the cam determines the particular nonlinearity.

On a dynamic basis, the results of the two methods of linearization differ, though they may be the same on a steady-state basis. Figure 11-12 demonstrates that the angularity method produces a flow vs control signal characteristic that is linear on both a steady-state and a dynamic basis. With a linear stroking time of the actuator, the time (0 to 100 percent) vs flow (0 to 100 per cent) characteristic is linear. In addition, the flow vs control signal characteristic is also linear.

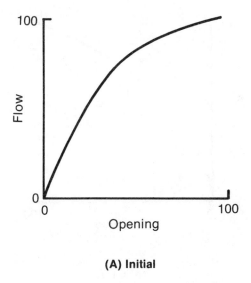

(A) Initial

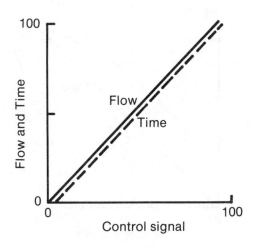

**(B) Linkage angularity plus
straight-line positioner**

Figure 11-12 Desired Damper Characteristics

In Figure 11-13, a cam positioner and parallel linkage are used. The steady-state characteristic of flow vs control signal is linear, but the time vs flow characteristic is nonlinear due to the nonlinear position vs flow characteristic of the actuator. This is important during very rapid changes in load when a proper match between fuel and air may be necessary on a second-by-second basis. This time linearity also improves the precision of tuning the control loop over the complete range of load.

The actuators for dampers are commonly called control drives and can be either pneumatic or electric as desired. They need not be matched to the operational medium or signal type of the control system. In Figure 11-14, several different ways of controlling a pneumatic piston actuator are shown.

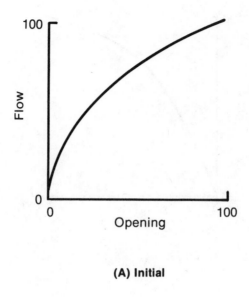

(A) Initial

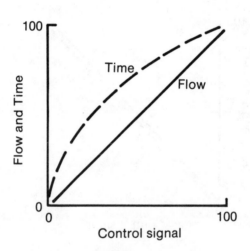

**(B) Straighten characteristic —
with positioner only**

Figure 11-13 Desired Damper Characteristics

If the control system is pneumatic, the signal is connected directly to the positioner of the piston. If the control signal is an electric analog signal such as 4 to 20 mA, then an I/P (current-to-pneumatic) converter is used to convert the signal to a proportional pneumatic signal. For an electric analog signal, the piston positioner can be changed to one that will receive the electric analog signal directly.

Similarly, electric digital pulse signals can be used to control a pneumatic control drive. One method is the use of a motor-operated pneumatic loader to convert the digital signal to a standard pneumatic signal with the pneumatic signal connected directly to a standard pneumatic control drive. Another method is the use of a digital positioner that will convert the digital pulses to a corresponding piston motion.

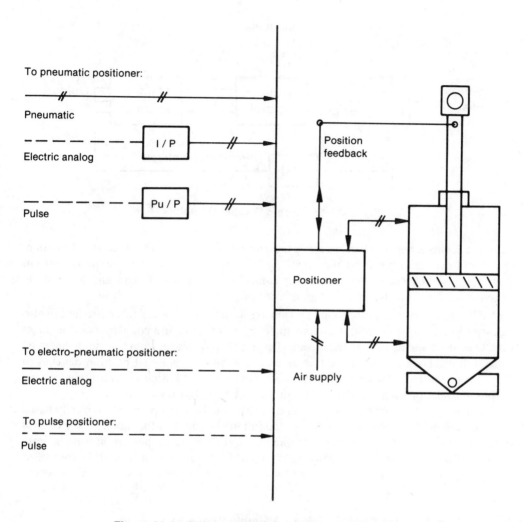

Figure 11-14 Pneumatic Piston with Optional Positioners

The torque rating of a pneumatic piston operator is usually based on a piston pressure differential of approximately 20 psi less than the full air supply pressure. Additional torque can be developed if the full air supply pressure is applied to one side of the piston with 0 pressure on the other. This is likely, however, to result in erratic positioning. To use a lower differential reduces the power rating of the device.

Pneumatic piston operators are not inherently self-locking on air supply failure. By using a compressed air receiver of sufficient capacity, they can operate for a period of time after an electric power failure that stops the air compressors. If the design engineer judges it to be necessary, mechanical locks can be applied to the control drives to lock them in position upon a failure of the air supply. It is normally not sufficient to merely trap the air that is in the piston. Air failure locks are costly and many engineers feel that the same amount of expenditure to improve the reliability of the supply air is more cost-effective.

Electric motor-operated damper control drives are usually inherently self-locking upon power failure. Many engineers feel that this action is a significant advantage for the use of electric rather than pneumatic operators. In these operators the rotary motion is converted to angular motion by various drive screw or worm gear arrangements.

A typical electrical arrangement is shown in Figure 11-15. A difference device measures the difference between the control signal and a characterized position feedback signal. Any

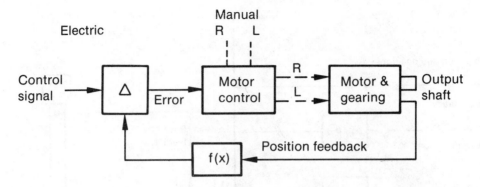

Figure 11-15 Electric Damper Operators

error that results is amplified and controls the current to the motor. The direction of rotation of the motor is determined by the polarity of the position error. To operate the motor manually, a switch disconnects the motor controller, and a set of raise and lower push buttons is used to increase or decrease the controller variable.

In the above case the motor follows an analog control signal. A digital controller can also be used to pulse drive the motor directly in either a clockwise or a counterclockwise direction. One of the inputs to the digital controller is a position signal from the motor-operated control drive. The controller uses this signal along with a computed position demand to direct the pulse-driven motor. Characterization can be obtained using a curve-fitting formula in the controller or by the feedback of a precharacterized position signal.

The maximum torque or power rating of an electric damper operator is typically based on approximately 25 percent over the normal full load current of the motor. If the control drive is loaded to exceed this value, a torque limit switch prevents the current from increasing. The 25 percent overload is usually required only for breakaway or to start the controlled device moving. It should not be designed for use in normal operation, since to do so would continually overload the motor.

11-6 Draft or Air Flow Control Using a Variable Speed Fan

If the only concern were the control of the draft or air flow, the fans would be operated at constant speed. Another concern, however, is the reduction in the power requirements of the fans.

A typical set of fan characteristics is shown in Figure 11-16. Referring to this figure, if it is assumed that the boiler is 100 per cent loaded and the flow requirement is 60,000 cfm, then the power requirement is approximately 87 shaft horsepower. If the load is reduced to a flow requirement of 50,000 cfm and fan speed remains at the 1160 rpm level, then the power requirement is approximately 85 shaft horsepower. For this approximately 17 percent load reduction, the power requirement has been reduced only approximately 2.3 percent.

If the fan speed is reduced to 960 rpm simultaneously with the load reduction, the power requirement is reduced to approximately 48 shaft horsepower. For this 17 percent reduction in flow, the power reduction is approximately 55 percent. Note that both sets of values are on the system resistance curve. In addition, the speed reduction is also approximately 17 percent. In all cases the percentage reduction in speed approximates the percentage reduction in flow capacity.

One of the laws of fan performance related to speed states that:

(1) capacity or cfm varies directly as the fan speed,
(2) pressure varies as the square of fan speed, and
(3) power varies as the cube of fan speed.

If the fan were sized exactly for full load at full speed, then a normal 4:1 flow control turndown would require a 4:1 speed turndown for the fan. In a normal installation it is not uncommon to find fans designed with considerable excess capacity. In such a case, the full load fan speed may be 50 to 70 percent of the design fan speed. The result is a fan speed turndown requirement of 6 to 8:1 for a capacity turndown of 4:1. The above assumes that the fan speed alone is used to control the capacity.

Fans can be operated under variable speed control to save fan power in several ways. If the fan is steam turbine driven, the speed can be modified by adjusting the speed setting of the speed governor or by simply using a control valve to control the steam flow to a turbine. If the fan is motor-operated, two-speed or variable speed motors may be used. Fan speed may also be varied, even though a constant speed motor is used, by using a variable speed magnetic or hydraulic coupling between the motor and the fan.

In whatever manner the fan speed is adjusted, the speed response oftens deteriorates rapidly at speeds below approximately 1/3 rated speed. This affects the air flow control dynamics with respect to fuel flow but may not affect speed positioning accuracy. Since a fan

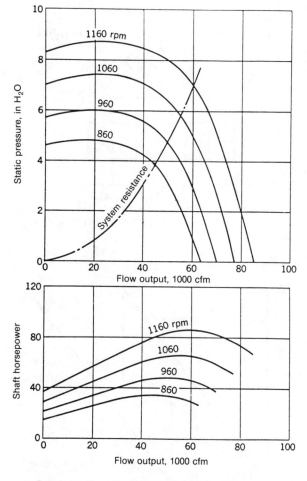

Graph to show how desired output and static pressure
can be obtained economically by varying fan speed to avoid
large throttling losses.

Figure 11-16 Variable Speed Fan Characteristic Curves

(From *Steam, Its Generation and Use,* ©Babcock and Wilcox)

in a typical installation is often designed with excess capacity, the capacity turndown at 1/3 rated speed is approximately 2:1, much less than that required. The net result is that most installations require some damper control participation at lower loads to extend the rangeability of good control response.

Figure 11-17 depicts a satisfactory non-interacting method of combining the damper and speed control. This is called *split range control*. The positioners of the actuators are adjusted to split the control signal range with overlap in the center of the range.

The damper control drive is adjusted to open the damper completely as the control signal changes from 0 percent to approximately 60 percent. The speed control device of the fan is adjusted to accelerate the fan speed from 1/3 rated speed to the desired full load speed as the control signal changes from approximately 40 percent to 100 percent. In the overlap area between 40 and 60 percent, the damper is becoming less responsive as it nears the 100 percent open position, and the fan speed is less responsive due to its being near the 1/3 rated speed. Combining the two in this portion of the range results in a somewhat uniform flow response over the entire range.

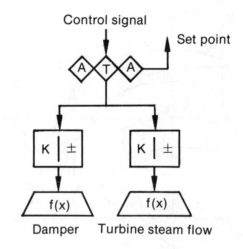

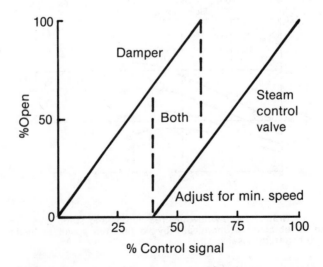

Figure 11-17 Combined Damper and Fan Speed Control

If the speed control mechanism is a steam turbine, experience has provided some lessons in the application. Some of the methods are listed below in the order of their suitability.

1. A separate conventional steam control valve ahead of the speed governor is non-interacting with the speed governor and is a good control solution. Particular care is necessary in sizing the valve, since the turbine power (analogous to turbine steam flow) may have a turndown of 10 or 15:1 as the control valve pressure drop increases over a range of 10:1. In some cases the clearance area around the valve plug may be too great for the desired minimum flow condition.

2. Using a device to change the speed set point on a high quality hydraulic or electronic speed governor is usually a satisfactory solution for a wide range control. It may be necessary to detune the flow control loop to avoid interaction between the flow and the speed control loops.

3. A control valve in the oil circuit of a hydraulic governor has the same potential interaction problems of method 2 above and in addition has a considerably narrower speed control range.

4. The simple application of a diaphragm operator to a mechanical speed governor is usually not satisfactory. The motion vs steam flow characteristic and the valve stroke of the speed control valve result in poor control if the valve is used as a flow control valve.

The basic rules for fan speed control with the end purpose of reducing the auxiliary power requirements of the fan drives are:

(1) Operate the fans at the lowest speed that is consistent with good control response.

(2) Operate control dampers as near open as possible consistent with good control response.

(3) Apply the damper control to the fan inlet rather than the discharge.

Figure 11-18 shows the relationship between power requirements for the various methods of motor-driven fan output control. The curve for a turbine drive fan or a variable frequency motor-driven fan would be similar to that shown for a magnetic or a hydraulic coupling. In all cases, as shown by Figure 11-18, inlet vane control consumes less power than discharge damper control.

11-7 Minimum Air Flow

Current National Fire Protection Association (NFPA) regulations tend to limit the air flow control range requirement to 4:1. This results from the recommended — 25 percent of full load air flow — as a minimum flow for many types of boilers. Some general statements can be made concerning this minimum:

(1) It does not apply to stoker-fired boilers or to any boilers that retain a significant fuel storage within the furnace.

(2) For a single-burner gas or oil fired boiler with a constant speed combustion air fan, the minimum limit can often be applied as a mechanical stop on the control damper.

(3) For a multiburner boiler, the minimum flow rate control should be based on flow measurement. This is necessary since changing the number of burners changes the resistance of the flow path.

(4) If the minimum air flow is 25 percent of full load air flow with 10 percent excess air (110 percent total air), reducing the fuel flow below 25 percent will result in a large increase in excess air. For example, reducing fuel to 15 percent of full load to satisfy load demand would result in 83 percent excess air — $((110 \times 25)/15) = 183$ percent total air.

(5) When it is known that the boiler steam flow range will be greater than 4:1, the control designer should take care to include necessary anti-windup features in the control system design.

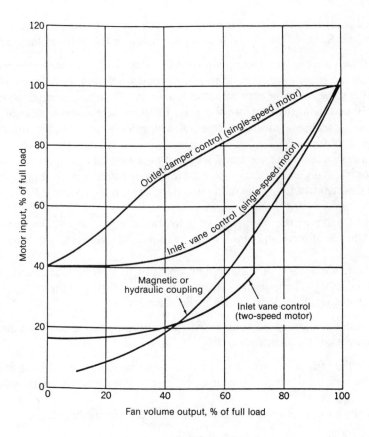

Figure 11-18 Comparison of Methods for Saving Fan Power

(From *Steam, Its Generation and Use*, ©Babcock and Wilcox)

Section 12
Measurement and Control of Furnace Draft

The need for the control of furnace draft occurs only in balanced draft boilers. In Section 11 it is stated that the normal practice is to control furnace draft at a very slightly negative pressure set point. When such a pressure measurement is made and used for control purposes, the stack effects must be carefully considered.

12-1 Measurement of Furnace Draft

Figure 12-1 demonstrates the choice of pressure tap locations for measuring furnace draft. The pressure connection on most boilers is located on the front, side, or roof of the furnace. Although the measurements at these three locations would be for the same furnace chamber of a particular boiler, the measurement values would differ due to the differing stack or chimney effects. The measurements at different elevations will differ by approximately 0.01 inches H_2O per foot elevation. The measurement in the roof of the furnace will be the highest value.

Since it is necessary to have negative pressure at all points, the value at the furnace roof becomes the controlling factor in determining the desired set point for the control of furnace draft. Thus, if the pressure at the furnace roof is to be minus 0.1 inch of H_2O and the connection for measuring furnace draft is located at an elevation 15 feet below the furnace

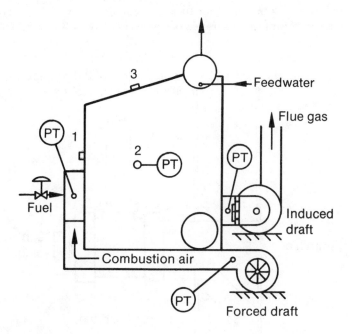

Notes:
1, 2, 3 — alternate furnace draft pressure connections.
Reading changes approx. 0.01 in. of H_2O per ft elevation.

Figure 12-1 Measurement of Furnace Draft

roof, then the set point for this control loop should be approximately -0.25 inch of H_2O. On a large boiler the connection might be as much as 50 feet or more below the roof elevation. In this case, the set point should be approximately -0.6 inches of H_2O or at a lower pressure.

Because of the very low pressure involved, the pressure connection should be large enough so that slight changes in the furnace draft can be very quickly felt by the measuring instrument. General practice is shown by Figure 12-2. The actual connection is 2-inch pipe size, and the piping to the instrument is ¾ to 1 inch in size. The 2-inch connection is provided with a tee and a plug in order that the plug can be removed and the connection easily cleaned.

In some cases involving balanced draft coal or solid fuel boilers, it is appropriate to drill a small hole (approximately ⅛ inch) in the plug. This allows a small amount of air to be drawn into the furnace at all times to help prevent soot or ash from plugging the connection. This procedure should never be used with pressure-fired boilers. For these boilers it is necessary that the instrument connection systems be free of all leaks in order to avoid the introduction of H_2O vapor, soot, or ash into the connecting piping.

For most boilers a furnace draft or pressure transmitter will operate normally within a pressure range of less than 1 inch of H_2O. For presenting the information to an operator, a normal instrument pressure range of +0.1 to −0.5 inch of H_2O is typically used. Such a narrow range is not normally satisfactory for control purposes. On fast changes of flow capacity or under abnormal operating conditions, the actual pressure or draft may exceed this range and thus not provide the controller with all the intelligence necessary during the period of change.

Furnace draft measurement is also subject to considerable process "noise". The use of a narrow range transmitter tends to accentuate the effect of such noise in the measurement. An additional factor is primarily a limitation of analog control. In this case it is quite often impossible to reduce the controller gain to a low enough value. The general practice is therefore to use a control transmitter range of approximately +1.0 to −5.0 inches of H_2O.

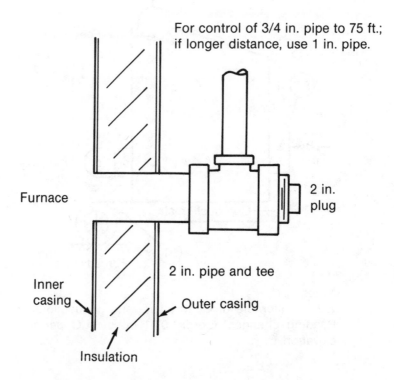

Figure 12-2 Measuring Tap for Furnace Draft

12-2 Furnace Draft Control Using Simple Feedback Control

The simplest form of the furnace draft control loop uses a simple feedback control loop. In this case the control of air flow is usually assigned to the forced draft with the furnace draft control regulating the level of induced draft. Generally, it is most desirable to measure air flow on the forced draft side of the furnace. Assigning the air flow control to forced draft tends to reduce interaction between the air flow and the furnace draft control loops.

The control arrangement is shown in Figure 12-3. As shown, the air flow capacity is changed by modulating the forced draft. The resulting change in furnace draft feeds back to the controller causing a *series* change to the induced draft. It is also possible to assign the air flow change to the induced draft with the series action taking place on the forced draft. In this case the controller action is reversed.

On many installations a control loop of this type is very difficult to tune for satisfactory results under dynamic load changing situations. The series action of the control allows too much time difference between the changes to the forced and induced drafts. Theoretically, these should be moving in parallel. In addition the large amount of process noise as a percentage of the measurement signal tends to require tuning adjustments of lower than desirable gain and slower than desired integral. In some cases, if feedback control alone is used, it may be necessary in order to achieve control loop stability to remove all proportional action and rely on integral control alone. This tends to accentuate the problem of the series time delay.

For comparison purposes, Figure 12-4 demonstrates the performance of a typical feedback furnace draft control loop. The excursions tend to be large with respect to the set point value, and the control tends to be unstable due to the effects of the process noise.

Such a control loop may be the single most difficult boiler control loop. Assuming a measurement at the furnace roof, the goal is to hold the furnace draft to a set point of

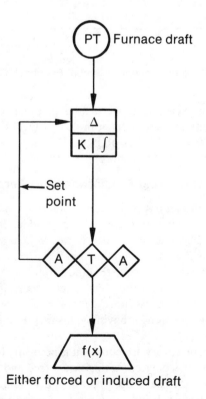

Either forced or induced draft

Figure 12-3 Furnace Draft Control (Single-Element Feedback Control)

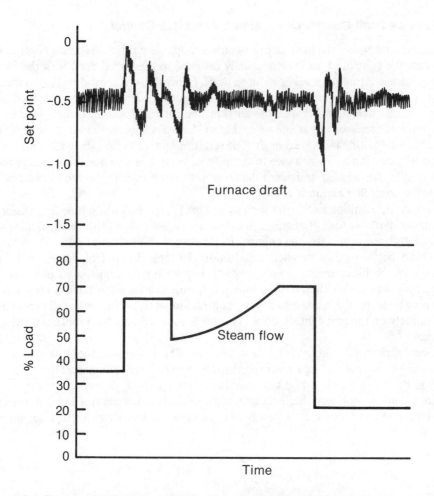

Figure 12-4 Furnace Draft Control (Single-Element Feedback Control of Induced Draft Fan)

−0.1 inch of H_2O with an excursion range of plus or minus 0.05 inch of H_2O, while the process noise is usually a minimum of plus or minus 0.1 inches of H_2O and a typical overall capability of the fans at 6 to 10 inches of H_2O. For large electric utility boilers the fan capability may be 25 inches of H_2O or more.

12-3 Furnace Draft Control Using Feedforward-plus-Feedback Control

Figure 12-5 demonstrates an improved control loop for the control of furnace draft. In this case the signal to the forced draft control device is added in the summer (item a) to the output of the furnace draft feedback controller. In this way the series time lag between forced and induced control action is eliminated. Note that it is necessary to provide a bias function in the summer (item a). This is necessary so that the output of the furnace draft controller will operate normally in the middle portion of its output range. This allows the controller to equally add or subtract from the feedforward signal as necessary. A proper control alignment for the summer (item a) would show it having gains of 1.0 on both inputs and with bias of −50 percent.

In applying this or other feedforward control it is necessary to parallel the flow characteristics of the two parallel control devices (in this case forced and induced drafts). If this is not done the two will not provide the proper parallel effect and much of the benefit of the feedforward control may be lost. It is also necessary to select the proper feedforward signal. If measured air flow is used as the feedforward signal, a series time lag is introduced. In

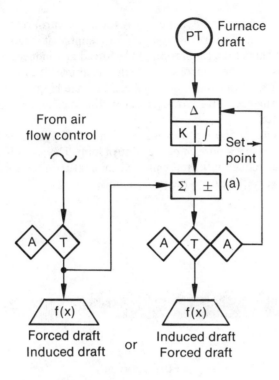

Figure 12-5 Furnace Draft Control (Feedforward-plus-Feedback Control)

addition, some positive feedback effect may be intoduced into the loop if there is any interaction between the air flow and the furnace draft measurement.

Figure 12-6 shows performance of the feedforward system on a comparative basis with the feedback arrangement. In this case the capacity changes can be made with much smaller deviations from the furnace draft control set point. Because of the feedforward action, the furnace draft controller can be considerably slower in action without reducing the effectiveness of the control loop. This adds control stability by reducing the gain and integral requirements and thus reducing the effect of process noise. Since the forced and induced drafts operate in parallel, any potential interaction between the forced and the induced draft control is significantly reduced.

12-4 Furnace Draft Control Using Push-Pull Feedforward-plus-Feedback Control

In the diagram shown in Figure 12-7, the feedback portion of the control loop is improved by applying it in a push-pull manner. The feedforward portion of the loop is identical to the feedforward portion of the system described in 12-5. The control signal from the air flow controller is used as an input to the summer (item a). The other input to this summer is the output of the furnace draft feedback controller. Properly aligned, both of these inputs would have a gain of 1.0. As before, a –50 percent bias is applied to the output of the summer (item a).

An additional function, difference (item b), uses the same two inputs as the summer (item a). When properly aligned, both inputs to the difference function (item b) have a gain of 1.0, and a bias of +50 percent is applied to its output. This arrangement provides improved dynamic performance by allowing the feedback controller to add to the induced fan control signal while simultaneously subtracting from the forced draft control signal.

   The system can thus adjust on a dynamic basis for any control result that differs from
that calibrated into the basic feedforward system. For example, it is more "forgiving" as to
the paralleling requirements of the calibration of the forced and induced control devices. In
the basic feedforward system, the control signal vs flow characteristics are used to match the
forced and induced drafts. If flow resistances change, the matching deteriorates and affects
the feedforward performance. This arrangement automatically compensates for these
changes in flow resistance.

   Improved performance of the feedforward portion of the system reduces further the
control demand on the feedback portion of the control loop. The result is improved control
stability through further reduction in the gain and integral requirements of the feedback
controller, and thus lower effects from the process noise.

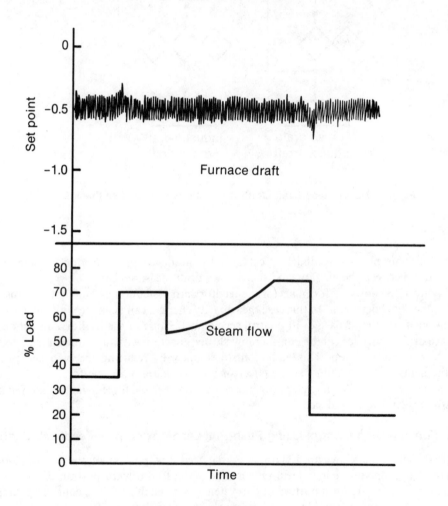

Figure 12-6 Furnace Draft Control
(Feedforward-plus-Feedback Control of an Induced Draft Fan)

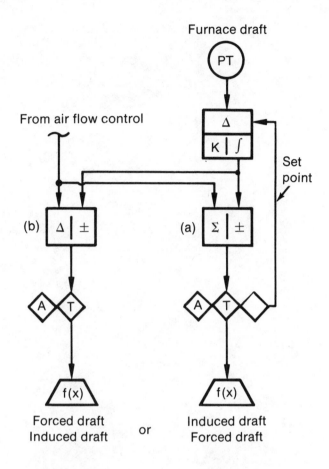

Figure 12-7 Furnace Draft Control
(Push-Pull Feedforward-plus-Feedback Control)

Section 13
Measurement and Control of Combustion

The measurement and control of combustion air flow is key to the proper functioning of any boiler combustion control system. Accurate measurement of the combustion air flow is difficult and often must be improvised into the boiler system design by the control designer. It is important to remember that relative air flow is much more necessary for controlling boiler performance than accuracy of weight or volume of air flow. The important consideration is the proper amount of combustion air to burn the fuel available at any particular point in time.

In view of this, the correct manner of calibrating the air flow measurement is to do so by the use of field combustion tests. By using such tests the air flow measurement is calibrated to match, on a relative basis, the fuel flow or other measure of air flow requirement. The field calibration compensates for the variation in required excess combustion air as the boiler load changes, plus any individual characteristic of the air flow primary measuring element. The net result is that the flow vs differential pressure calibration of this measurement is rarely a true square root relationship even though it normally approximates a square root.

13-1 Measurement of Air Flow

Combustion air flow is customarily measured with some form of primary measuring element that is installed as a part of the boiler duct and fan system. This is used with a differential pressure measurement device. The ducts are of various shapes and sizes; they also have numerous 90 degree bends, short straight runs, and other features that are normally considered to be detriments to accurate measurement. These factors have a very significant effect on the actual flow coefficients and their characteristics of flow vs differential pressure. This is one factor that necessitates field calibration by using the results of boiler combustion tests.

Any permanent pressure drop in the system as a result of the installation of the primary element increases the requirement for power to drive the combustion air fans. For this reason it is desirable that the primary element have a low differential pressure at full boiler capacity. Typically, the secondary differential pressure measuring devices have design differentials of 1 to 2 inches of water at maximum signal output.

Different types of primary elements have different discharge coefficients. The result is a difference in permanent pressure loss. The choice between primary elements based on permanent pressure loss (and thus fan power consumption) is often difficult to justify on an economic basis. Consider that the difference might be that of discharge coefficients of 0.6 and 0.85. If the full load differential pressure is 1 inch of H_2O, the permanent pressure loss would differ by 0.25 inch of water at full load. This would, however, be reduced to 0.0625 inch of H_2O at 50 percent load and 0.0156 inch of H_2O at 25 percent load.

One potential primary device is an orifice segment in the forced draft duct. Figure 13-1 shows this type of device. It is simple to design and install, but its drawback is lower pressure recovery and thus greater permanent pressure drop. Considering the individual nature of the ductwork, an accurate design is impossible. An approximate design combined with field calibration produces good results.

An approximate design can be made by considering the duct as a round duct and designing an orifice plate in a standard manner. The d/D (orifice diameter/pipe diameter) is then converted to an area ratio (a/A), which will be the square of the d/D ratio. Using the area ratio, the opening area can be determined. This area is subtracted from the duct cross-section area to yield the area of the orifice segment.

In order to reduce the permanent pressure loss of the measuring device, a venturi-type duct segment, as shown by Figure 13-2, can be installed. The design of such a duct segment should only be undertaken by someone with good design basis information, such as a boiler manufacturer. This does not assure a good design, however, since the author experienced one case in which a design for 2 inches of H_2O differential yielded an actual differential pressure of 8 inches of H_2O. A recalculation confirmed the original design.

Further reduction in permanent pressure loss can be obtained by using an air foil design as shown in Figure 13-3. The design of an air foil also requires background of such a design along with empirical data that is based on the actual results of previous air foil designs. Air foil designs are usually made by boiler manufacturers. A primary device of this type is also somewhat less expensive to construct than the venturi duct section.

Other differential pressure primary element devices that can be used are various devices based on the pitot principle. In the pitot tube, the pressure differential is the difference between the static pressure and the velocity head or pressure. Such devices are the pitot

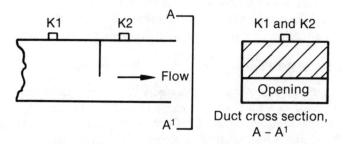

Note: K1 and K2 are pressure connections for ΔP.

Figure 13-1 Measurement of Combustion Air Flow with Orifice Segment in Duct

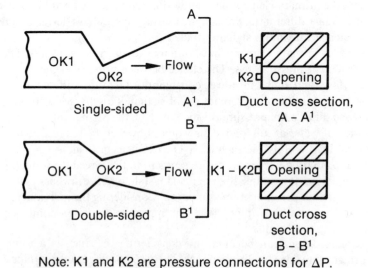

Note: K1 and K2 are pressure connections for ΔP.

Figure 13-2 Measurement of Combustion Air Flow with Venturi Section in Duct

venturi, the piezometer ring, the "piccolo" tube, the Annubar™, and other forms of the pitot tube. In some cases these are used in multiples in order to obtain averages of different points within the duct. For these devices the permanent pressure loss is very small and thus some power saving results.

Another technique that requires no additional power consumption is to the use the pressure drop across the boiler parts. One method is the use of the pressure drop across the air side of a tubular air preheater as shown in Figure 13-4. There are usually 2 or more inches of H_2O available at full boiler load. In most such air preheater arrangements, the difference in elevation between the pressure connections requires compensation for the chimney or stack effect due to the difference in temperatures. The method of connection shown in Figure 13-4 will usually provide the necessary compensation. Using the preheater pressure drop is not a satisfactory method with a rotary regenerative air preheater because of variable flow path cleanliness and variable seal leakage.

Since the combustion air accounts for over 90 percent of the mass of the flue gas products of combustion, a measurement of flue gas flow can be used as an inferential measurement of

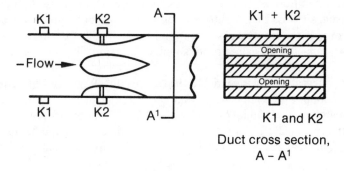

Note: K1 and K2 are pressure connections for ΔP.

Figure 13-3 Measurement of Combustion Air Flow with Air Foil in Duct

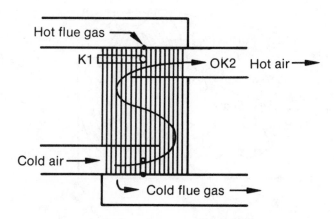

Note: K1 and K2 are pressure connections for ΔP.

**Figure 13-4 Measurement of Combustion Air Flow
Using Differential Pressure across a Tubular Air Preheater**

combustion air flow. Figure 13-5 shows this method, which uses the pressure or draft differential across the boiler tube passes. The use of such a measurement tends, however, to produce a greater interaction between the fuel and air flow control loops. A further disadvantage is that such an air flow measurement is affected by soot or other foreign deposits on the boiler tubes. Another disadvantage is the unavailabilty in many cases of sufficient draft loss. As shown here, a difference in elevation of the pressure connections is used to compensate for the "chimney" effect that results from temperature difference of the flue gases at the two measurement points.

13-2 Control of Air Flow

Either open-loop or closed-loop control can be used for air flow control. An example of these two control arrangements is shown in Figure 13-6. In the open-loop arrangement the combustion air flow demand resulting from the boiler steam load is satisfied by positioning the controlled device. If the driving force ahead of the controlled device changes or system flow resistance changes, the open-loop arrangement will allow air flow to change.

To compensate for such changes, closed-loop feedback control is used. In this case, a deviation from the air flow set point feeds back to reposition the controlled device in order to maintain a given air flow. The controller shown utilizes both proportional and integral control functions. If the flow measurement and the controlled device are reasonably well matched in flow capacity, the gain (proportional) of the controller can usually be set between 0.5 and 1.0.

The correct integral setting is geared to the total feedback time (usually a few seconds) of the flow control loop. The result is usually an integral (repeats per minute) setting of several rpm. The gain and integral tuning of the loop is also affected by process noise. It should be remembered that the air flow control time response ultimately should be matched with fuel flow response. This may result in one of these loops having less than optimum tuning.

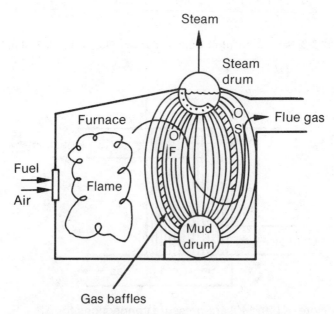

Note: F and S are pressure connections for ΔP.

**Figure 13-5 Inferred Measurement of Combustion Air Flow
by Pressure Drop across Baffles in Flue Gas Stream**

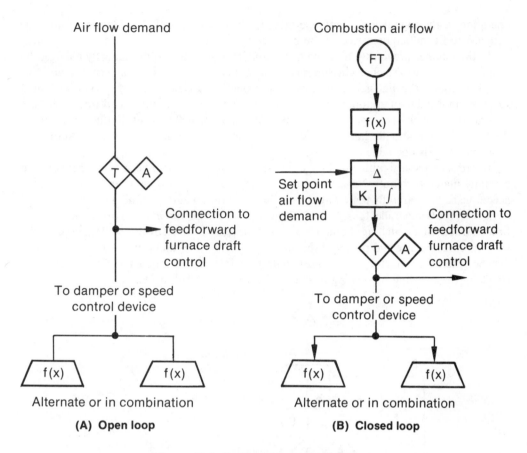

Figure 13-6 Combustion Air Flow Control

If the boiler uses both forced and induced draft fans, it is desirable to connect the control signal to the controlled device as the feedforward signal in a feedforward-plus-feedback furnace draft control loop. This tends to reduce or eliminate interaction between the air flow and furnace draft control loops.

The arrangement above concerns installation with not more than one forced draft fan or one set of forced and induced draft fans. If two or more fans normally operate in parallel to supply combustion air, then the single-fan failure mode must be considered. If two fans operate in parallel, the failure of either fan would allow the output of the operating fan to be lost through the openings to fan suction of the non-operating fan. In general, the requirements for parallel fan systems are:

(1) A change in gain between 1- and 2- or more fan operation.

(2) Automatic closing of shutoff dampers on the inoperative fan to avoid air recirculation.

(3) The ability to balance the fan loads.

(4) Usually, additional control devices on the fan discharge dampers in order to achieve tight shutoff of air flow.

(5) Opening all dampers with both fans tripped.

All of the above are achieved in the control arrangement shown in Figure 13-7. In the case of a single fan trip, digital logic operates the transfer switch (a) in the control circuit of that fan and also operates the common transfer switch (b). In this way the 0 percent signal (e) is connected to the controlled devices on the fan that has tripped. Should the second fan trip, the switches (a) will be in their tripped condition but the common switch (b) will switch admitting the 100 percent control signal (f) to both sets of control drives. The key to the operation is the digital logic that operates the switches (a) and (b). This logic must be designed to fit the requirements of the particular installation.

In the manner previously described, the control loop gain is automatically changed by summing the control signals in summer (c) and balancing the sum against the air flow demand signal in the high gain-fast integral controller (d). In this case with two fans of equal size the input gains of summer (c) would be 0.5. The shutoff damper control devices are calibrated for quick opening when the control signal is above 0 percent. In this arrangement the fans can be manually balanced by using the manual bias controls shown. Automatic balancing can also be added if desired.

In order to avoid significant changes in air flow when a fan trip occurs, it is necessary to carefully match the control characteristics of the two fans that operate in parallel. This is accomplished by carefully matching the air flow calibration and the timing of the controlled devices on the two parallel fans. If the boiler uses both forced and induced draft, the time and flow characteristic matching should involve all four fans (2 forced draft and 2 induced draft). In some cases the size and power of the damper control drives and thus their stroking speed may be different for the forced and induced drafts. Matching the characteristics in these cases results in the speeding or slowing of one of the sets of control drives.

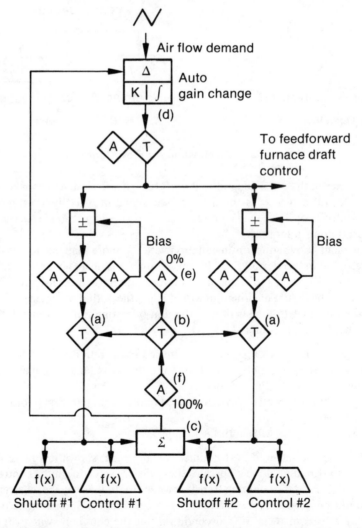

Figure 13-7 Air Flow Control (One or Two Fans in Parallel, Open Loop)

Section 14
Draft and Air Flow Control-Related Functions

Some of the boiler operation functions are closely related to the control of draft and air flow. These are the control of flue gas dew point and the blowing of soot from the boiler tubes.

14-1 Flue Gas Dew Point Control

This type of control is necessary only in cases where heat recovery equipment, such as an air preheater or an economizer, is applied to a boiler. In these cases the flue gases are cooled to a lower temperature than with a simple boiler with no heat recovery. Both of these heat recovery devices can be represented by the diagram shown in Figure 14-1 of a counterflow heat exchanger. In the case of an economizer, the incoming fluid would normally be in excess of 220° F. In the case of an air preheater, the incoming combustion air might normally be less than 100° F.

In order to avoid corrosion at the cold end of an air preheater, it is necessary to maintain the flue gas temperature above the dew point temperature. The flue gas dew point temperature is determined by the moisture content of the flue gas and the presence and percentage of SO_2 and SO_3 in the flue gas. From the standpoint of moisture alone, the lowest expected dew point temperature would be approximately 135° F for flue gas from natural gas with no sulphur content. The addition of even small amounts of sulphur in the fuel, and thus SO_2 and SO_3 in the flue gas, causes a significant shift upward in the dew point temperature. The resulting moisture would be a weak solution of sulphurous and sulphuric acid.

Natural gas is normally free of sulphur though some, called sour natural gas, contains sulphur in the form of hydrogen sulphide. There is often some small percentage of sulphur in fuel oil whose flue gas has a lower water vapor content than natural gas. Coal may also have a significant sulphur percentage, but its flue gases have a lower moisture percentage. Thus fuel oil, for a given sulphur percentage, may produce flue gas with the highest dew point temperature, often in excess of 200° F.

At the "cold end" of an air preheater, if the air temperature is 80° F and the flue gas temperature is 300° F, the average metal temperature in contact with the flue gas is considered to be 190° F, the average of the flue gas and combustion air temperatures. With a dew

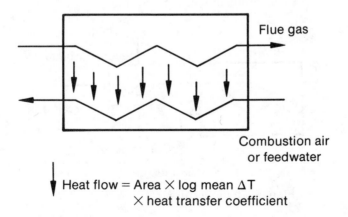

Flue gas

Combustion air
or feedwater

Heat flow = Area × log mean ΔT
× heat transfer coefficient

Figure 14-1 Economizer or Air Preheater Counterflow Heat Exchanger

point temperature of 200° F, acid moisture would collect on the metal surface and corrosion would take place.

This can be avoided through use of a control method that raises the incoming combustion air temperature so that the average cold end metal temperature is above the dew point. As the incoming air temperature is raised, the flue gas temperature shifts upward a similar amount. In the example above, raising the incoming air temperature to 120° F would cause flue gas temperature to shift upward to approximately 340° F. The average metal temperature would then be 230° F, well above a 200° F dew point temperature. If the dew point temperature were 230° F, then the combustion air temperature would have to be raised still higher than the 120° F point.

A control method that accomplishes this is shown in Figure 14-2. An air heater, with steam as the heating medium, is placed in the combustion air stream ahead of the flue gas heat recovery air preheater. The steam is controlled to this heater in order to develop the desired combustion air temperature. The control loop controlling the steam flow is shown in Figure 14-3. A simple feedback control loop as shown is usually adequate. The average of the flue gas and air temperatures, shown as the feedback, is considered a pseudo metal surface temperature.

If corrosion-resistant materials are used, the allowable metal temperature can be decreased. This basic problem can also affect the corrosion of metal stacks when they are used. In these cases, the problem occurs due to low outside ambient temperatures. The solution to this problem is the use of corrosion-resistant materials for the inside of the stacks or the insulation of the stacks.

The operation of this type of dew point control has a small impact on air flow and draft control by changing flow resistance on both the flue gas and air sides of the boiler. A 40° F rise in air preheater inlet air temperature will change the average specific volume across the air side of the preheater and change its pressure drop by approximately 2.5 percent. The change in draft loss on the flue gas side will be a smaller percentage amount. This change will be only slightly noticeable even though open-loop air flow control is used.

Other types of dew point control rely on variable recirculation of a portion of the preheated air back to the forced draft fan inlet. This method has a greater impact on the air flow control by requiring a variation in the opening of the inlet vanes of the forced draft fan.

14-2 Soot Blowing

Soot buildup on the boiler tubes is a normal occurrence for liquid or solid fuel boilers. This soot accumulation can be reduced by maintaining the correct combustion conditions. In

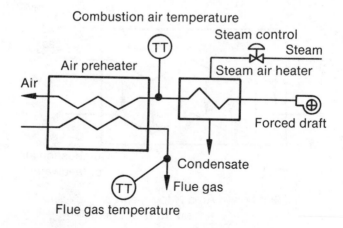

Figure 14-2 Dew Point Control Process Arrangement

any event it must be removed periodically in order to avoid a severe loss in heat transfer. Figure 14-4 illustrates the loss in heat conductivity due to soot accumulation. Soot is an excellent insulator, as shown, and a thin layer can significantly reduce heat transfer. The effect is an increase in boiler draft loss and flue gas temperature, with a resultant loss in boiler efficiency.

Soot is normally removed from the tubes with devices called soot blowers. These are devices mounted along the sides of the boiler from which jets of steam or compressed air are used to blast the soot from the tubes. Some fuels have chemical characteristics that cause the soot to adhere to the tubes. In these cases fuel additives are often used to change the characteristics of the soot in order that it may be more easily removed from the tubes.

The normal practice is to start the soot blowing at the furnace and sequentially blow soot into the flue gas stream. Operation of soot blowers near the front of the boiler, in particular, may cause severe pulsations in the furnace draft. For this reason it may be desirable to reduce the set point of the furnace draft control and increase the level of combustion air flow during soot blowing periods.

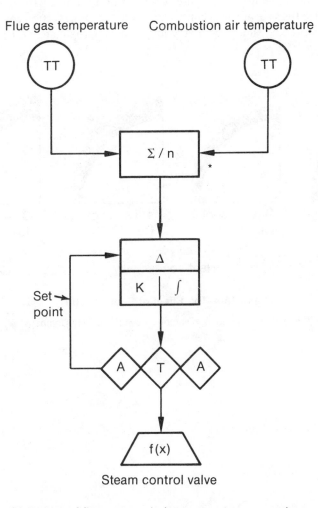

*Average of flue gas and air temperatures equals
pseudo "cold end" metal temperature.

Figure 14-3 Flue Gas Dew Point Control

Since the 1960s efforts have been made to direct the soot blowing sequence through the use of computer control. The success of such schemes depends on the availability in the computer of a relatively sophisticated boiler model along with the input of continuous measurements of boiler excess air, flue gas temperatures, boiler tube temperatures, and other factors. In this way local dirty spots in the boiler may be cleaned in a random sequence as required by the model. In addition to improved heat transfer, a claimed benefit of computer-directed soot blowing is a reduction in the cost of the compressed air or steam blowing medium.

Some more simplistic methods of computer-directed sequencing of soot blowers based on measurements of only boiler draft losses and flue gas temperatures have been investigated. The overall problem is quite complex and simple solutions may be no more economically effective than the traditional time-based sequencing.

Soot blowing may also impact the feedwater control performance. In some cases saturated steam for the soot blowers is taken directly from the steam drum output and is unmeasured. In these cases the measured steam flow and the measured feedwater flow of a three-element feedwater control system will not match. The effect is to cause a small shift in the boiler drum level set point. If this causes a serious problem in a particular case, integral action can be added to the drum level controller. As stated in the section covering feedwater control the integral setting should be a very low value, approximately 0.1 rpm (repeat per minute).

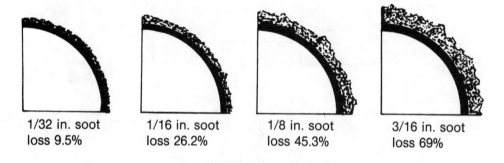

| 1/32 in. soot loss 9.5% | 1/16 in. soot loss 26.2% | 1/8 in. soot loss 45.3% | 3/16 in. soot loss 69% |

Heat conductivity

Figure 14-4 The Effect of Soot Buildup

(From a technical paper, ©Diamond Power Specialty Company)

Section 15
Gaseous and Liquid Fuel Burners

All burners must perform five functions:

    (1) Deliver fuel to the combustion chamber
    (2) Deliver air to the combustion chamber
    (3) Mix the fuel and air
    (4) Ignite and burn the mixture
    (5) Remove the products of combustion

Regardless of the type of fuel used, the burner must perform all five of these functions. In the case of coal or oil, delivering fuel to the combustion chamber also includes preparing the fuel so that it will burn. Section 5 discussed how gas and oil were delivered to the burning system and how oil and coal are prepared for burning. Figure 5-6 shows that the fuel must be gasified, and this is part of the first function. Figure 5-6 also shows that turbulence is one of the three "T"s of combustion.

The source of the energy that produces this turbulent mixture is the kinetic energy of the fuel and air streams. Restrictions in each of these streams at the burner convert potential energy of the fuel and air, due to their pressures, into kinetic energy. This is done by restricting the flow areas and thus increasing the fuel and air velocities as they travel through the burner. In reducing fuel and air flow rates as the burner is turned down, this "mixing energy" is sharply reduced.

The result is that there is a normal requirement for higher excess air as the flow rate of a given burner is reduced. Figure 15-1 shows the typical relationship between excess air and flow rate for gaseous (including pulverized coal) or liquid fuel-burning systems. Combustion control systems must automatically compensate for this relationship by shifting the fuel-air ratio as the burners are turned down.

15-1 Burners for Gaseous Fuel

Figure 5-1 shows the simplicity with which gaseous fuel is automatically delivered to the plant as it is used. The characteristics of the burner design must be adjusted to match the characteristics of the fuel delivered to the burner by the firing rate control. Pressure at the burner is one means of classifying gas burners. This classification can be described as follows:

    (1) Low pressure — 2 to 8 ounces per square inch
    (2) Intermediate pressure — 8 ounces per square inch to 2 psig
    (3) High pressure — 2 to 50 psig

Burners are selected according to the particular gaseous fuel and the pressure available, or to fit the flame characteristics desired. The gas issues from the burner through small orifices that vary in size and capacity depending on the maximum pressure provided. The higher pressures usually have a greater *turndown* capability. Turndown is the firing rate ratio over which the firing rate can be reduced without flame instability.

Combustion air may enter the burner through atmospheric pressure, which is higher than the suction at the burner. Such a burner is called an atmospheric burner. The suction is either developed by natural draft or by inspiration in which a relatively high velocity jet of gas creates suction as it enters the combustion chamber. Mechanical draft, created by a forced draft fan or an induced draft fan, or both, is the other method of delivering air to the burner. In order to control the fuel-air ratio accurately and for burner safety, all the combustion air should enter through or around the burner.

The most common burners in use in boilers are those that are sometimes referred to as nozzle mix burners. The fuel issues from a small orifice and is injected into and thoroughly mixed with a whirling vortex of air. A counterrotation whirl is often imparted to the issuing gas in order to provide greater turbulence. The basic idea is to mix the fuel and air thoroughly so that complete combustion will take place with a minimum of excess air.

A small pilot burner is usually used to obtain ignition. This pilot can be continuously lit, or it can be ignited each time the burner is started, and shut down after the burner is lit. Ignition characteristics often are affected by the type and shape of the flame and whether the fuel and air are mixed immediately or in stages. Staged combustion is obtained by initiating the flame with primary air and later admitting the rest of the air that is necessary to complete the combustion. Staged combustion produces a longer and lower temperature flame and fewer nitrous oxides.

The final function of the burner is to remove the products of combustion. This is accomplished by their replacement with new mixtures of fuel and air that are forced into the combustion chamber by fans and fuel pressure.

In Figure 15-2, a cross section of a ring-type gas burner is shown. This is a common type of high-pressure gas burner for boiler use. The normal pressure operating range is from approximately 1 to 10 psig. As the air passes through the air registers the velocity is increased and a whirl is imparted to the air, which then is forced toward the burner throat. The gas issues from the small holes shown on the inside of the hollow gas ring and is injected by the fuel pressure perpendicular to the air stream. When such burners are properly adjusted, they can be operated at a minimum pressure of approximately 1 psig while maintaining a stable

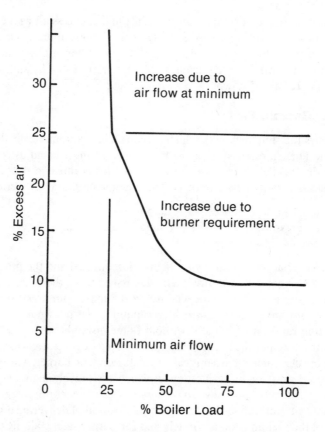

Figure 15-1 Excess Air vs Boiler Load, Typical Curve

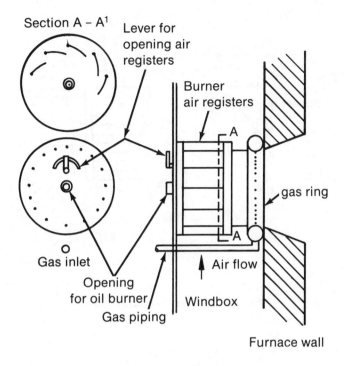

Section A – A¹

Lever for opening air registers

Burner air registers

A

gas ring

Gas inlet

Opening for oil burner

Air flow

Windbox

Gas piping

Furnace wall

Figure 15-2 Ring-Type Gas Burners

flame. Because of the relatively small gas orifices, dirt or foreign substances in the gas stream may tend to plug the burners. The boiler must be shut down to clean such burners.

Figure 15-3 depicts a "gun-type" gas burner. In this burner the air register, with its function of imparting a whirl to the combustion air, is similar to that of the ring burner. The gas, however, is injected into the air stream from the center instead of the outer periphery. Because space is more limited on the tip of the smaller diameter gun, the gas orifices are larger and fewer in number than on the ring burner. An advantage of this burner over the ring burner is that the gas orifices are larger and need less cleaning; they can, however, be cleaned by removing the gun while the boiler is in operation. When properly adjusted, burners of this type can be operated at pressures in excess of 20 psig and with a minimum pressure stable flame at approximately 1 psig.

Figure 15-4 represents a spud-type burner. In this burner the gas ring is external, with a number of smaller guns or spuds connected to the ring. This burner is a design for obtaining the benefits of both the ring and the gun burners. The spuds can be removed for cleaning while the boiler is in operation, and the gas orifices are greater in number and smaller for better dispersal of the gas into the air stream.

The burners above are all of the nozzle mix type. Other gas burners for boilers are designed so that the gas and air issue from the burner in somewhat parallel streams. The fuel-air mixing takes place in the furnace instead of at the burner.

Figure 15-5 shows a burner of this type. Although only oil and coal nozzles are shown, gas nozzles can be added to or replace such oil and coal nozzles. As shown in Figure 15-5 these burners are mounted at the corners of a furnace and directed so that the gas and air streams are tangent to a vortex or "fireball" in the center of the furnace. Since the fuel and air mix gradually, the combustion process is slower and at lower temperature. The result is that such burners naturally produce lower NOx (nitrous oxides) in the flue gases.

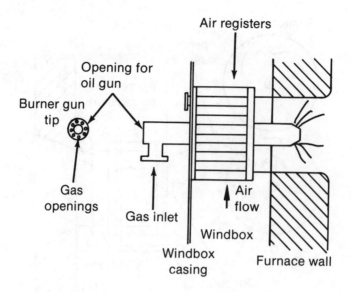

Figure 15-3 Gun-Type Gas Burner

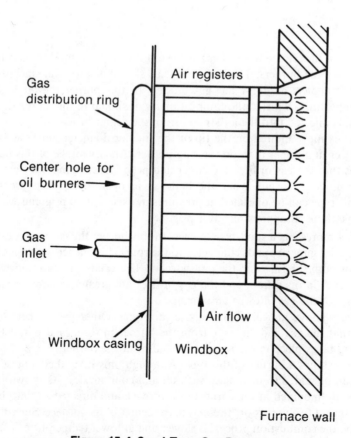

Figure 15-4 Spud-Type Gas Burner

The turndown of a gas burner is related to its maximum pressure and the minimum pressure that will support a stable flame. The turndown ratio can be calculated if the minimum and maximum pressures are known. If the maximum pressure produces flow in the subcritical (below Mach 1.0) range, the burner can be considered as a flow orifice with flow varying in accordance with the square root of the differential pressure. The burner pressure is the upstream pressure, and the downstream pressure is atmospheric pressure. The pressure at the burner changes the flowing density, which affects the flow in accordance with the square root of the absolute pressure ratio. If 14.7 psia is considered the base, the

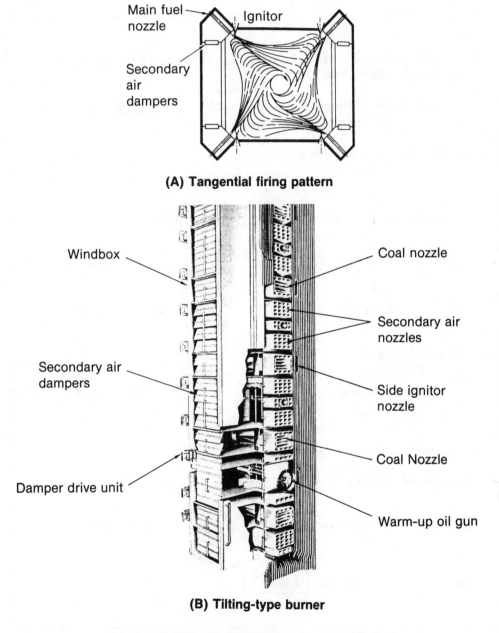

(A) Tangential firing pattern

(B) Tilting-type burner

Figure 15-5 Tilting Tangential-Type for Gas, Oil, Coal

(From *Fossil Power Systems*, © Combustion Engineering Inc.)

$$\dot{m} = \zeta A \sqrt{2\rho \Delta p} = \zeta A \sqrt{2 \frac{p}{RT} \Delta p} \quad \propto \sqrt{\rho \Delta p} \quad \text{if} \quad T = C$$

174 $\dfrac{\dot{m}_8}{\dot{m}_1} = \sqrt{\dfrac{(14.7+8)8}{(14.7+1)1}} = \sqrt{\dfrac{22.7}{15.7} \cdot \dfrac{8}{1}} = 3.401$ *The Control of Boilers*

approximate turndown can be calculated in accordance with Figure 15-6A. Note that a very slight reduction in the minimum pressure significantly increases the turndown ratio.

For higher pressures producing flow in the critical velocity range (greater than Mach 1.0), the flow varies as a relationship with a combination of differential pressure and with the burner absolute pressure. Figure 15-6B is an example of the calculation of burner turndown for a maximum burner pressure of 25 psig. In the critical velocity range, flow change is directly proportional to change in the absolute inlet pressure. For this burner, above approximately 70 percent of maximum flow the velocity is critical and below this point it is subcritical.

- Depends on burner design for maximum pressure to carry the load and minimum for stable fire.

- Consider burner an orifice — basic $\sqrt{}$ relation pressure (pressure drop to flow) but density changes considerably.

- Plot burner flow curve min to max

Example

ASSUME   1 psi minimum, 8 psi maximum — 100%

$$\text{7 psi flow} = \sqrt{\frac{21.7}{22.7}} \times \sqrt{\frac{7}{8}} \times 1.00 = 0.91455$$

$$\text{6 psi flow} = \sqrt{\frac{20.7}{22.7}} \times \sqrt{\frac{6}{8}} \times 1.00 = 0.827$$

continuing down to

$$\text{1 psi flow} = \sqrt{\frac{15.7}{22.7}} \times \sqrt{\frac{1}{8}} \times 1.00 = 0.288$$

Turndown   = 1/0.288   or $\boxed{\text{3.47 to 1}}$

If minimum stable fire is reduced by 0.5 psi to 0.5 psi,

$$\text{then 0.5 psig flow} = \sqrt{\frac{15.2}{22.7}} \times \sqrt{\frac{0.5}{8}} \times 1.00 = 0.2046$$

Turndown   = 1/0.2046 or $\boxed{\text{4.89 to 1}}$

Figure 15-6A Turndown of Gas Burners (Subcritical)

For all the above burners, the firing rate is adjusted by regulating the fuel and combustion air in parallel. The control device for the fuel is a control valve in the fuel supply line as shown in Figure 5-1. All of the above burners also show an opening in the center for insertion of a fuel oil gun so that the burner may be used as a combination oil and gas burner.

15-2 Pulverized Coal Burners

Though coal is a solid fuel, in pulverized coal systems where coal is finely ground and transported to the furnace in a primary air stream, some burners are similar to the gun-type

---

- Basic — Follow subcritical curve to 0.47 of absolute inlet pressure, then straight line to maximum pressure.

Example

ASSUME   Atmosphere = 14.7 psia,
          pressure drop to atmosphere = 0.47 of absolute pressure at X psig.

$$0.47 \ (X + 14.7) = X$$
$$0.47X + 6.909 = X$$

$$0.53X = 6.909$$

$$X = 13.03 \text{ psi for critical velocity}$$

ASSUME   25 psig maximum, 1 psig minimum.

25 psig = 100% or 1.0

13 psig = 27.7/39.7 = 0.698

$$1 \text{ psig} = \sqrt{\frac{15.7}{27.7}} \times \sqrt{\frac{1}{13}} \times 0.698 = 0.145$$

Turndown = 1/0.145 or $\boxed{6.90 \text{ to } 1}$

- If minimum can safely be reduced to 0.5 psig,

$$0.5 \text{ psig} = \sqrt{\frac{15.2}{27.7}} \times \sqrt{\frac{0.5}{13}} \times 0.698 = 0.1014$$

Turndown = 1/0.1014 or $\boxed{9.86 \text{ to } 1}$

Figure 15-6B Turndown of Gas Burners (Critical Velocity)

gas burners. They have similar air registers, often with provision for auxiliary oil or gas fuel to be burned in the same burner. Figure 15-7A is a drawing of such a combination burner. Figure 15-7B shows the detail of two modern pulverized coal burners. One of these is a staged combustion burner with secondary and tertiary air registers that allow the shaping of the flame and introduction and mixing of the fuel and air in stages. The benefit is a reduction of NOx in the flue gases.

A low NOx burner from another manufacturer is shown in Figure 15-7C. Note that this burner is also provided with dual air registers for use in mixing the fuel and air in stages.

Figure 15-5, which was shown previously in connection with gas burners, shows a tangential burner with nozzles arranged for coal and with provision for a warm-up oil gun. Secondary air which is added to the primary air to complete the combustion process issues from separate air nozzles in streams that are parallel to the coal/primary air streams.

In both of these types of burners the pulverized coal supply to the boiler is changed by adjusting the speed of the coal feeder that admits coal to the pulverizer and by changing the primary air flow rate. A change in primary air flow immediately affects the flow of that coal which is already ground and stored in the pulverizer. Refer to Figure 5-5 for the coal feeding and furnace supply arrangement.

15-3   Fuel Oil Burners

Fuel oil burners for boilers usually use the same air register mechanisms as are used for gas burners. An oil gun is inserted into the center of the burner ssembly. The oil gun atomizes and sprays the oil outward into the whirling combustion air stream. Though some gas burners do not reguire what are known as "diffusers" or "impellers", These devices are necessary for oil and pulverized coal burners and larger gas burners (Figure 15-8).

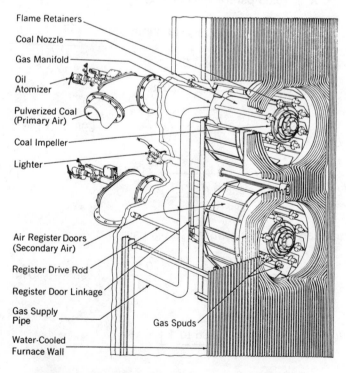

Figure 15-7A Pulverized Coal Combination Burner

(From *Steam, Its Generation and Use,* ©Babcock and Wilcox)

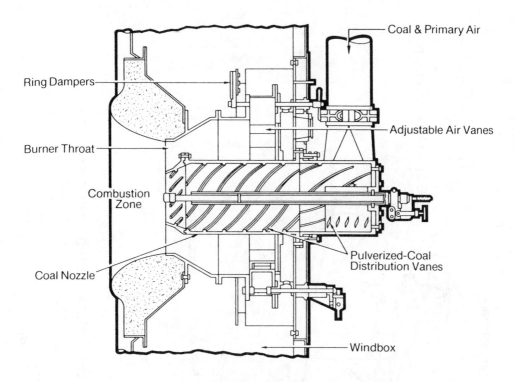

Horizontal Firing Pulverized Coal Burner

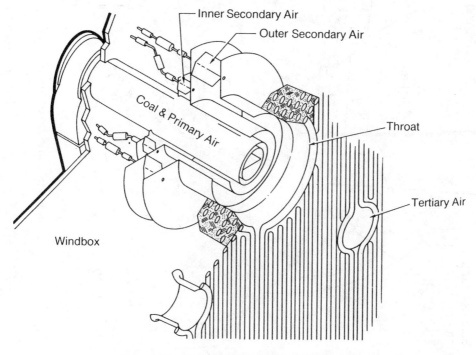

Staged Combustion Pulverized Coal Burner

Figure 15-7B Pulverized Coal Burners

(From *Fossil Power Systems*, ©Combustion Engineering Co. Inc.)

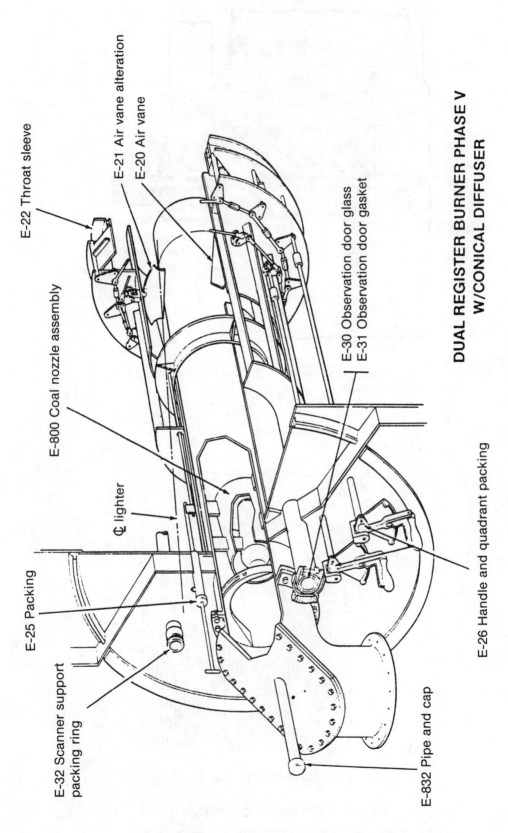

E-22 Throat sleeve

E-21 Air vane alteration
E-20 Air vane

E-800 Coal nozzle assembly

℄ lighter

E-25 Packing

E-32 Scanner support
packing ring

E-832 Pipe and cap

E-30 Observation door glass
E-31 Observation door gasket

E-26 Handle and quadrant packing

**DUAL REGISTER BURNER PHASE V
W/CONICAL DIFFUSER**

Figure 15-7C Low NOx Pulverized Coal Burner

(From *Babcock and Wilcox Co.*)

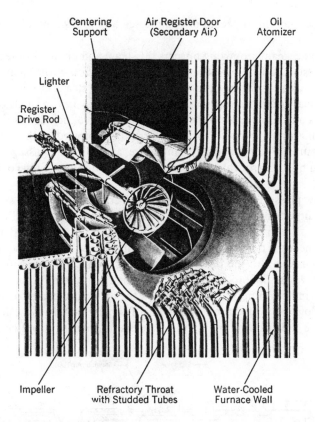

Figure 15-8 Impeller (Diffuser) for Gas, Oil, Pulverized Coal Burners

(From *Steam, Its Generation and Use*, ©Babcock & Wilcox Co.)

The purpose of the diffuser (impeller) is to produce more stable ignition by splitting the combustion air stream into primary and secondary portions. A portion of the air passes through the diffuser slots and creates a fuel rich zone as the primary air mixes with the fuel issuing from the burner gun. The remainder of the combustion air flows around the outer edge of the impeller and is added as secondary air to the basic fuel/primary air mixture.

The oil guns for different burners are similar but may differ in the atomizing method. The basic classification is:

(1) Steam or compressed air atomizing
(2) Mechanical pressure atomizing
(3) Mechanical return flow pressure atomizing

The steam or compressed air atomizing burner uses the mechanical energy of steam or compressed air to atomize the oil and is not dependent upon the thermal content of the atomizing fluid. In view of this, air and steam may be substituted for each other in the same burner. As shown in Figure 15-9, the steam and oil arrive at the burner tip through concentric tubes. The jets of oil and steam (or air) merge at the tip and the oil stream is atomized. The oil supply is regulated by a control valve in the supply line. Atomizing steam or air is regulated with a differential pressure-regulating valve that controls the steam or air at a given set pressure above the burner oil pressure.

A typical burner gun tip for steam or compressed air atomization of fuel oil is shown in Figure 15-10A.

Caution: In some steam atomizing burners the steam and oil mix inside the gun. The use of differential pressure control for atomizing steam on such a burner may make the flame unstable. In those cases it may be necessary to control the steam in parallel with the fuel.

In the mechanical pressure atomizing burner, the gun may have a single oil tube to the tip or it may have a concentric return flow line. The tip orifice includes tangential slots that cause the oil to whirl as it sprays from the burner and is atomized by the potential energy of the oil supply pressure. When the burner is in service, any return oil passage is closed, and oil supply is regulated by a control valve in the supply line. A tip for this type of burner is shown in Figure 15-10B.

The return flow pressure atomizing burner can operate at the highest supply pressure of all oil burners. It utilizes concentric tubes in the gun, with the supply oil flow in the outer tube and the return flow in the center tube. If the return flow is unrestricted, the oil flows to a whirling chamber near the tip, whirls, and by centrifugal force enters the return flow passage instead of being sprayed into the furnace. Restricting the return flow with a control valve causes a greater or lesser flow to be sprayed from the tip into the furnace.

Since oil flows in both supply and return lines, the oil burned can be determined only by measuring both flows and subtracting the return from the supply flow. It is necessary to carefully match the flow calibrations of these meters in order to avoid large errors in "oil

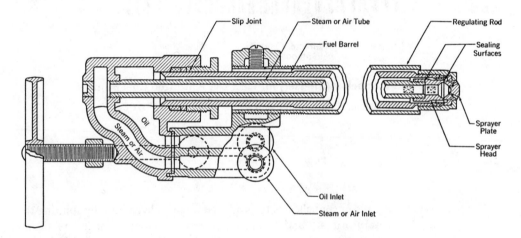

Figure 15-9 Steam (or Air) Oil Atomizer Assembly

From *Steam, Its Geration and Use*, ©Babcock and Wilcox Co.)

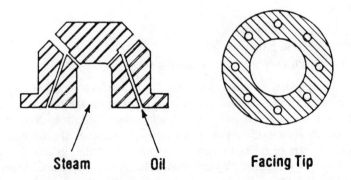

Figure 15-10A Steam (or Air) Atomizing Y Jet

burned" at low furnace input rates. Figure 15-11 shows a cross section of a typical burner tip for a return flow type of burner. Also shown are the flow characteristic curves that relate supply and return flows to the pressures involved.

Figure 15-12 is a chart showing application guidelines and other information for these and other fuel oil burners as a group.

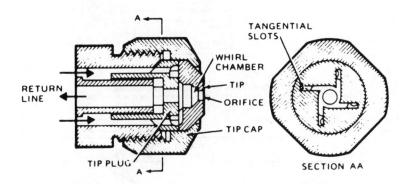

Figure 15-10B Pressure Atomizing Oil Burner Tip

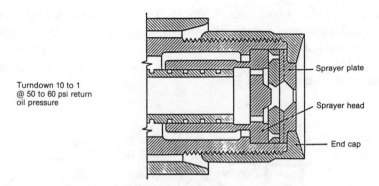

Turndown 10 to 1
@ 50 to 60 psi return
oil pressure

(A) Mechanical return flow oil atomizer detail at furnace end of atomizer assembly showing sprayer head, sprayer plate, and end cap.

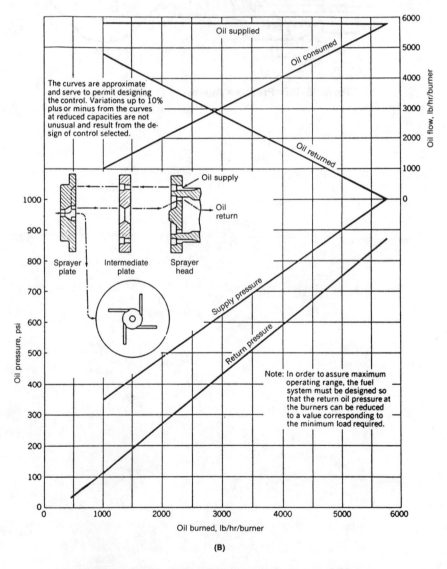

(B)

Figure 15-11 Return Flow, Mechanical-Atomizing Oil Burner

(From *Steam, Its Generation and Use,* © Babcock and Wilcox)

Type of burner	Approximate size (input)		Atomizing pressure (PSIG)		Turn-down ratio[b]	Usual applications
	Gal Per hr	BTUH[a]	Oil	Air or steam		
Vaporizing	0.15 to 2.5	20 thousand to 350 thousand	—	—	2 to 1	Small capacity residential stoves, furnaces, and water heaters.
Air atomizing, low pressure	0.5 to 530	70 thousand to 80 million	—	½ to 2	3 to 1 up to 8 to 1	Most versatile; warm air furnaces, boilers, and process furnaces.
Air atomizing, high pressure	10 to 500	1.4 million to 75 million	—	25 to 150	3 to 1 up to 8 to 1	Major-sized industrial plants utilizing compressed air in their process applications. Particularly adaptable for converting to combination gas-oil burners.
Steam atomizing	10 to 500	1.4 million to 75 million	—	25 to 150	3 to 1 up to 8 to 1	Major-sized industrial plants using steam generators, particularly water tube boilers. Particularly adaptable for converting to combination gas-oil burners.
Mechanical atomizing, nonrecirculating	0.5 to 80	70 thousand to 12 million	75 to 300	—	2 to 1 (on-off control only)	Domestic warm air furnaces, boilers, and small industrial furnaces.
Mechanical atomizing, recirculating (return flow)	25 to 1200	3.5 million to 180 million	100 to 1000	—	3 to 1 up to 10 to 1	Most economical atomizing burner. Wide range, from domestic oil burners to major-sized boiler plants, including marine boilers.
Horizontal rotary	5 to 300	750 thousand to 45 million	—	—	4 to 1	All types of installations — domestic, industrial, and commerical.
Vertical rotary	0.3 to 15	40 thousand to 2 million	—	—	4 to 1	Residential and small industrial burners.

[a]British thermal units per hour, based on a heat content of 140,000 Btu per gal for light (distillate) oils, and 150,000 Btu for gal for heavy (residual) oils.

[b]Ratio of the maximum firing rate to the minimum firing rate at which the burner will operate satisfactorily.

Figure 15-12 Oil Burner Application

(From *Steam, Its Generation and Use*, ©Babcock and Wilcox)

Section 16
Flue Gas Analysis Trimming of Combustion Control Systems

In Section 6, the effect of flue gas constituent percentage on boiler efficiency was discussed. It follows that flue gas analysis can be used to precisely control the ratio of fuel to air.

There is a general perception that flue gas analysis instrumentation is more complex, more costly, and less reliable than other types of measuring instruments normally used in boiler control systems. On the other hand, the analysis of flue gas provides a measurement that is more precise than fuel/air ratio measurements that use simpler methods. Flue gas analysis does, however, consume some time, is after the fact, and the combustion process can change very rapidly. In order to achieve control precision along with timeliness of control action and maximum reliability, the general practice is to design the basic control system using the most reliable basic instrumentation and to use flue gas analysis as a superimposed trimming control loop.

Flue gas analysis trimming control is applicable to combustion control systems for all types of fuel firing and all types of basic boiler control systems from the simplest to the most complex. Since the application is universal, it is being discussed separately from the discussion of the more individual basic systems.

There are a number of variations of the analyzer based trimming control loops. These variations are based primarily on the particular flue gas analysis made. There are also degrees of desired fuel/air ratio precision, control sophistication, and costs as a byproduct of these variations. In this context the desired fuel/air ratio is that which results in the lowest cost operation of the process.

16-1 Useful Flue Gas Analyses

There are several flue gas analyses that are potentially useful in combustion control trimming loops. The analyses are used individually or in combination, based on the type of trim control desired. Generally, the analyses and the combinations in which they may be used are as follows:

Analyses

(1) % Oxygen (% O_2) — Excess combustion air is a function of % oxygen.

(2) % Opacity — A measurement of smoke or particulate matter. Due to environmental standards, this measurement may be needed for use in limiting the control action based upon other analyses.

(3) % Carbon Dioxide (% CO_2) — When total combustion air is greater than 100% of that theoretically required, excess combustion air is a function of % carbon dioxide.

(4) Carbon Monoxide (CO) or Total Combustible in the PPM Range — This measurement is that of unburned gases. Measurement in the ppm range is necessary if desired control precision is to be obtained.

Uses

(5) % Oxygen as an Individual Control Index. (This is the most tested, with control application since the early 1940's)

(6) PPM CO or Total Combustible as an Individual Control Index. (Application of this method began approximately 1973.)

(7) % Oxygen in Combination with PPM CO or Total Combustible. (Applications began approximately 1977.)

(8) % Carbon Dioxide in Combination with PPM CO. (Applications began approximately 1977.)

(9) % Oxygen in Combination with % Opacity. (Applications began approximately 1977.)

The time of actual field application is important since the number of installations, the testing and feedback time period, and potential number of different types of applications on the multitude of boiler variations are important factors in solidifying the application practice.

16-2 Methods of Flue Gas Analysis

Since approximately 1970 new methods have significantly improved the reliability and precision of flue gas analysis. Prior to this time flue gases were analyzed by drawing a sample of the flue gas from the boiler ducts, washing and cooling the sample, and passing it into the flue gas analyzer. This action reduced the sample temperature below the dew point, and the water formed in the combustion process was condensed. Since the sample as analyzed did not contain this water, it was called a "dry" sample.

The sampling system also introduced a time lag in the analysis. This time lag was generally in the range of 30 seconds to 2 minutes. In addition, due to the potential for sample line plugging as the sample reached the dew point, the sampling system was the major source of maintenance in the analyzer system.

The newer generation of analyzers has made a quantum jump in reliability by analyzing the flue gas on the "wet" basis. This is done by using methods that do not remove the hot flue gas from the ducts to cool it and do not reduce the temperature below the dew point. There are three basic methods for accomplishing this.

16-2-1 The "In Situ" Point Sample Method

This method uses an analyzer probe, which is inserted into the duct at the point of analysis. An analysis cell on the end of the probe analyzes the hot flue gas flowing past it. With no sampling system time lag, typical response times of these analyzers is 5 to 10 seconds. Figure 16-1 shows the arrangement for this method. % Oxygen and ppm CO can be measured in this manner.

The analysis will be that existing at the point of cell location in the duct. If averages are needed, the outputs of multiple analyzers are averaged. The location along the flue gas stream is very flexible, needing only an open area on one side of the boiler duct system for probe insertion. Because of the air seal leakage of regenerative air preheaters, the analyzer probes should be installed in the flue gas stream ahead of such air preheaters.

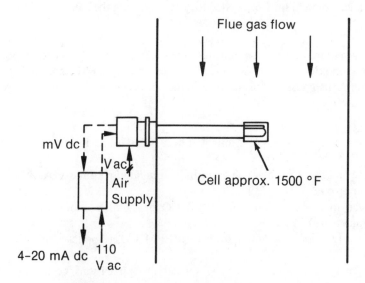

Figure 16-1 In Situ Flue Gas Analysis (Point Sample)

If % oxygen is being measured, the measuring cell is zirconium oxide. The principle is that of a fuel cell. The cell reacts to the ratio of the partial pressure of oxygen in the flue gas to the partial pressure of oxygen in reference air that is also admitted to the cell. Because of the cell temperature, residual combustible gas is burned in the cell and absorbs some of the oxygen in the flue gas. The result is therefore a measurement of % net oxygen.

The output of the cell is a millivolt signal that is an analog of the logarithm of the partial pressure ratio of the % oxygen in the flue gas to that of the reference air. In order that results will be accurate and repeatable, the cell temperature is closely controlled. The typical controlled temperature set point for such cells is in the 1300 to 1600° F range. The upper temperature limit for this method is that of the cell temperature.

The cell output (approximately 0 to 100 mV) is further processed by inversion, linearization, and amplification to produce a linear signal of % oxygen vs milliamps (mA). In this form the signal can easily be used by any form of standard control instrumentation.

If ppm CO is being measured, the analysis cell operates on the infrared absorption principle. In this principle carbon monoxide absorbs a part of the infrared energy from an included infrared source. The amount absorbed is within the frequency range specific to carbon monoxide. Measuring the absorption of infrared energy within that range provides a measurement of carbon monoxide.

The resulting electrical signal in some (not all) of the ppm CO analyzers is compensated for % moisture, temperature, and excess air and is processed and amplified to that of a standard instrumentation signal. The upper limit on flue gas temperature for this method is typically in the 600 to 700° F range. Above this range infrared energy emitted from the flue gas tends to interfere with that emitted by the infrared source. Some of the most recent designs have been compensated to allow an upper limit temperature of 1000° F.

16-2-2 Extractive or "Ex Situ" Method

The extractive or ex situ method is shown in Figure 16-2. In this method a small sample of the flue gas is drawn from the duct to the heated cell housing mounted on the duct wall. Because the sample is never cooled below the dew point and admitted back into the duct after analysis, sampling maintenance is not normally a problem. As with the "in situ" method, the analysis is on the "wet" basis.

For an air aspirated sample, the response time of this method is typically in the range of 10 to 15 seconds. For natural or thermal aspiration, the response time is in the range of 25 to 40 seconds. Both times consist of the sample transport time from the tip of the sample probe to the cell. The measurement is the analysis of the flue gas existing at the tip of the sample probe. If averages are needed, special averaging sample probes can be used or multiple analyzers can be used and their output signals averaged.

As with the in situ analyzers, the location along the flue gas stream is flexible since installation is made from one side of the boiler or flue gas ducts. Also as with the in situ analyzers, the installation should be ahead of sources of air input such as the leakage of the seals of regenerative air preheaters. This method is applicable to the measurement of % oxygen and ppm or % total combustible.

If % oxygen is being measured, the measurement principle is the same as that of the in situ analyzers. In this case the temperature controlled zirconium oxide cell (approximately 1500° F) is located in the cell housing, mounted on the duct instead of in the flue gas stream. The measurement signal and signal processing functions are also the same for the two methods.

Since the temperature-controlled cell is located outside the flue gas stream, cell temperature does not limit the application of this type of analyzer. In this case, the limit is based on the material of the sample probe. By using sample probes of ceramic material, flue gases can be analyzed with this system up to more than 3000° F.

If ppm total combustible is being measured, a catalytic combustion principle is used. In this case, any remaining combustible gas that is present in the flue gas is "burned" in the measurement cell to produce a signal that is then amplified to that of standard measuring instrumentation. The transport of the flue gas sample to the cell is identical to that of the % oxygen measurement. The cell can also be housed in the same assembly with the cell for measuring % oxygen and the same flue gas sample can be used. In this case, two output signals are obtained: % oxygen and ppm or % total combustible.

16-2-3 Light or Infrared Beam across the Stack

The third general method is the use of a beam of light or infrared energy across a flue gas duct or stack. This arrangement is shown in Figure 16-3.

The beam is transmitted through the flue gas from one side of the stack to the other. Both single-pass systems (measurement on the opposite side of the duct or stack from the source) and double-pass (in which the beam is reflected back to the side of the source) systems are used. Carbon monoxide in the ppm range and opacity are generally measured in this manner.

If ppm CO is being measured, the beam is infrared energy. The same principle of selective infrared absorption as used in the "in situ" ppm CO analyzer is also used in this type of analyzer. Analyzing the infrared absorption in an additional frequency range is sometimes used to measure % carbon dioxide (CO_2). As with the in situ measurement of ppm CO the normal application limitation of 600 to 700° F flue gas temperature applies. As indicated previously some manufacturers now specify 1000° F maximum.

If % opacity is to be measured, the essential principle is that a beam of light is shone across the duct. The measurement of light intensity that is received on the opposite side is a measure of % opacity.

For the applications described above, in general, locations must be found that are free of obstruction on both sides of the duct or flue gas stack. This often dictates that the analyzer be located in the flue gas stream after the air infiltration or leakage at the air preheater. This also tends to make the installation and location of this type of analysis less flexible and more costly

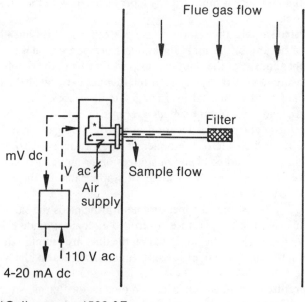

Figure 16-2 Extractive or Ex Situ Flue Gas Analysis (Point Sample)

than that of the single-side installations. While the above is correct in general, there is one design of this type of ppm CO analyzer that uses a single-side boiler duct penetration.

16-2-4 For and Against Factors

There are a number of different manufacturers who can supply analyzers as described above. Each manufacturer has designed his instrument based on his analysis of all the "for" and "against" factors. "For" and "against" cases can be made to favor particular choices of measurement or useful combinations of measurements. For example:

(1) In situ vs ex situ
(2) Point sample vs average sample
(3) % Oxygen vs CO ppm
(4) Total combustible ppm vs carbon monoxide ppm

(5) % Oxygen-CO vs % oxygen-opacity
(6) % Carbon dioxide-CO vs % oxygen-CO
(7) % Oxygen-CO vs % oxygen-total combustible

16-3 Pros and Cons of Measurement Methods and Gases Selected for Measurement

Various claims are made for the superiority of one measurement method over another. One comparison pits the "across the stack" measurement against the in situ and ex situ methods. In this case, the issue is the need for an "average" analysis as obtained with the "across the stack" method or the "point" analysis of the in situ or ex situ method.

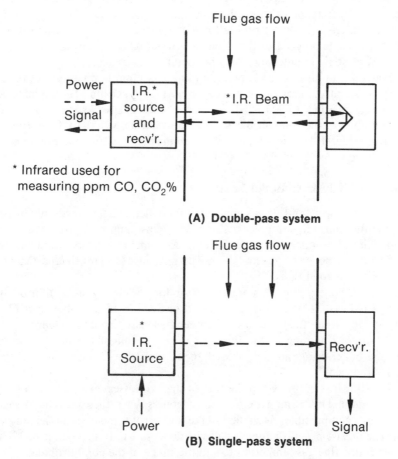

Figure 16-3 "Across the Stack" Flue Gas Measurement

The argument for "average" is that the total is measured and that an averaged analysis is a true measurement of all the flue gas. The argument for a "point" sample is that any air infiltration tends to flow along the sides of the boiler or duct where the air enters. Therefore, a point near the center of the duct is more indicative of burner performance.

Another comparison is made based on the need for "one-side" or "two-side" duct penetration. The mounting of the measuring device is much simpler and more flexible as to location for a single-side device. The relative location inflexibility of the "two-side" analyzer often requires its location downstream in the duct system from air leakage points such as those of the seals of regenerative air preheaters. Dilution of the flue gas with air changes all volumetric measurements of the flue gas constituents.

A "method" comparison is also made between the in situ method and the ex situ method. Response time is slightly shorter with the in situ method. Since the measuring cells of these analyzers have a millisecond range response time, in the in situ device an electronic design time constant is used to reduce the process noise of the measuremnt. In the ex situ device the sample transport produces a slightly longer time constant. The ex situ device requires a clean compressed air source for aspirating the sample. The quantity of compressed air is approximately 0.5 cfm. The in situ device uses a similar amount of clean compressed air in the cell as a reference % oxygen.

Other comparisons are based on ease of and cost of measuring cell replacement. The measuring cells of both devices tend to deteriorate with time. Since the cell is outside the duct in the ex situ device, a claim is made that cell replacement is simpler. The cost of the replacement cell is usually less for the ex situ device.

Opposed to the above, the in situ design appears to have some advantage for "dirty" fuels. For these cases, depending on the installations and characteristics of the soot or particulate matter, sample filters and sample probes for the ex situ device (non-existent in the in situ) may gradually become plugged. If this occurs, then periodic blow back may be required or other forms of periodic maintenance may be required.

A comparison can also be made based on the maximum temperature of the flue gases that can be measured. As previously stated, the normal upper limit for the infrared "across the stack" analyzers is in the 600 to 700° F range. For the in situ % oxygen device, the upper limit is the set point of the controlled cell temperature, usually in the 1300 to 1600° F range. For the ex situ analyzer the upper limit is based on the temperature limit of the sample probe used and is in the range of 3000 to 3200° F.

16-4 Flue Gas Analysis vs Boiler Load

The most desirable set of flue gas analyses for a boiler is determined only by the performance of the fuel burning equipment. Control systems cannot improve the basic performance capability of a boiler or its fuel burning equipment. A proper control system can, however, operate the boiler very near its "best" performance level for that particular load and other environmental conditions.

In general, all boilers require more excess air for complete combustion of the fuel as boiler load is reduced. This is briefly discussed in Section 15 and shown in Figure 15-1. Figure 16-4 demonstrates the relationship between fuel losses from excess air and those from combustible gases remaining in the flue gases. As a result of the increase in excess air requirement as boiler firing rate is decreased, the combustible gas curve shifts to the right as boiler load is reduced.

The lowest fuel loss occurs when the sum of the two losses (excess air and combustible gas) is at minimum. This point occurs on most boilers when the excess air is such as that which produces a combustible gas content of the flue gas at some point in the range of 200 to 2000 ppm. The precise point is a function of several factors that include the fuel burned, the excess air level, the flue gas temperature, and the shape of the combustible gas fuel loss vs excess air curve.

The following two basic control application theories are used for flue gas analysis trimming. Additional control approaches are based on combinations of the two.

(1) % Oxygen (% O_2) control is based on controlling to a fixed percentage of excess air based on a feedback of % O_2. Achieving the minimum fuel loss is accomplished by testing the boiler, determining the capabilities of boiler and fuel-burning equipment, and programming the % O_2 set point to match the test conditions. The desired set point is basically a function of the boiler firing rate. Usually, a small set point increase is added. This excess air "cushion" of approximately 3 to 4% is normally sufficient to allow for variations in the boiler-burner performance from that of the "as tested" condition.

(2) Control from ppm combustible gas is based on the assumption that, for different boiler loads and "best" economic performance, this measurement is at a relatively constant level. Figure 16-4 shows that when the combustible gas content of the flue gases is held to a constant value the excess air is automatically shifted as boiler load is changed.

If the goal is the optimum minimum fuel loss, the control requirement initially indicated and that has been generally used is a fixed set point feedback control from ppm combustible gas. It can be shown, however, that due to the characteristic of higher flue gas temperature at higher boiler load and with no change in the combustible gas loss vs excess air curve shape, there should be some shift in the set point with respect to boiler load, excess air level, and fuel burned.

In addition, data from operating plants has shown that the shape of the combustible gas loss vs excess air curve changes as boiler load is changed. The combustible gas fuel loss for each increment of excess air is a derivative of this curve. The combustible gas set point for an optimum minimum fuel loss must, therefore, be determined at each load by actual boiler test.

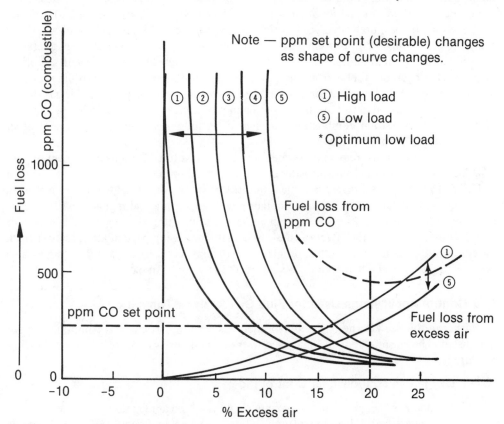

Figure 16-4 Excess Air-Combustible Gas Relationships

The results should then be used to develop a load-programmed set point for % combustible gas, just as is done for % oxygen.

Since any change in boiler-burner performance will automatically produce more or less combustible gas, no excess air cushion is required as with % O_2 control. The result is an incremental benefit in fuel loss reduction that results from the basic 3 to 4% excess air advantage for this approach to combustion control trimming.

16-5 PPM CO vs PPM Total Combustible Gas

Parts per million (ppm) combustible gas and ppm carbon monoxide (ppm CO) are not necessarily synonymous terms. While carbon monoxide is the most widely recognized combustible gas, the particular fuel-burning process and its combustion chemistry progression often produce other combustible residual gases "instead of" or "in addition to" carbon monoxide.

While the fact of other combustible gases has been proven in many boiler tests, there is insufficient data on the subject to state, without any reservation, the precise implication to control design. The result is a lack of total assurance that the governing gas should be carbon monoxide instead of the some other combustible gas such as one of the aldehydes or vice versa. A simple qualitative test for the presence of aldehydes requires only the bubbling of a small sample of flue gas through an aldehyde detector. The indications are that both CO and aldehydes are present to some degree.

A general guide is that knowledge of the flame characteristics can be useful in helping to define the application. A luminous flame, such as that of coal, fuel oil, and some gaseous fuel flames, indicates that carbon monoxide should probably be the controlling gas. A clear flame, such as that often obtained with rapid combustion of natural gas, may indicate that more aldehydes than carbon monoxide will be produced. Some of the evidence also indicates that, for such flames, preheated combustion air may increase the probability of aldehydes being the predominant combustible gas. The simple aldehyde test can confirm its presence.

Since a clear or nonluminous flame is obtained in practically all cases only with gas firing, then for all nongaseous fuels carbon monoxide should probably be the controlling residual combustible gas. It follows that for gas firing the correct controlling combustible gas should be carefully determined ahead of time. If this cannot be done or if the aldehyde test confirms its presence, then a total combustible gas sensor (including carbon monoxide and other combustible gases) should be used.

In some cases a measurement of % opacity may be substituted for a measurement of ppm CO. A comparison of the characteristics of % opacity and ppm CO vs % excess air is shown in Figure 16-5. Since % opacity is a measurement of smoke produced, its use as a control limit may be necessary to avoid noncompliance with environmental regulations. In stoker-fired boiler applications, % opacity may increase when operating above particular excess air levels. This results from the effect of flue gas velocity increasing particulate carryover from the combustion chamber. The effect of increasing opacity as excess air is increased may also occur with oil burners when high excess air results in a white smoke.

16-6 Control Applications Used for Flue Gas Analysis Trimming

Earlier in this section the basic control methods for flue gas analysis trimming control were listed. The specific control logic used in these methods is described in the following paragraphs.

16-6-1 Oxygen as an Individual Control Index

This method has been used for approximately 40 years, so application guidelines are well known and understood. This control application is shown in Figure 16-6. The function generator (a) develops a % oxygen set point signal as a function of boiler load or other index of firing rate. The particular function generated should be based on boiler tests at three or more boiler loads. The operator has the ability through the manually generated set point bias signal and summer (b) to shift the % O_2 set point curve up or down without changing its shape.

The controller (c) is tuned for a low gain and relatively slow integral response in order to obtain control stability. The low gain results from the relationship between total air flow change and % oxygen change. A change of 0.1 % oxygen is equal to approximately 0.5 % change in toal air flow. If the analyzer has a total range of 0 to 5% oxygen, then 0.1% O_2 is 2% of the total range. Since this relates at full load to 0.5 % of air flow range, then the gain limit is approximately 0.005/0.02 or 0.25.

The slow integral tuning of the controller results from the accumulated time constants in the control loop. These consist of controller and control tubing time constants (if pneumatic control), the time constant of the controlled device, the transport time from the control dampers and valves through the combustion process to the analyzer, and the analyzer time

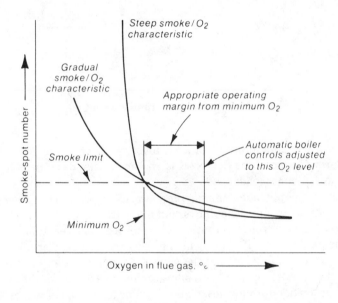

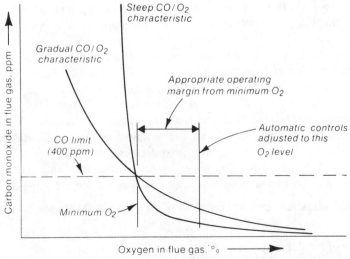

Figure 16-5 Oxygen/CO and Oxygen/Smoke Characteristic Curves

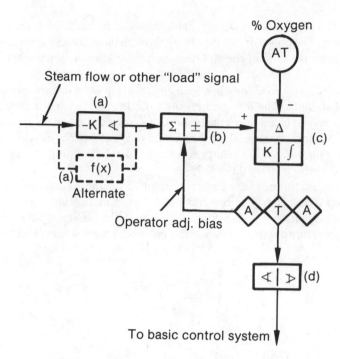

Figure 16-6 % Oxygen Trim Control Loop

constant. Under the best conditions, all of these together may be 15 to 20 seconds or more. Of these, the transport time of the flue gases is an almost pure delay time constant, which varies as a function of boiler load.

Because of the greater potential for analyzer failure as opposed to the generally used flow measurements, limits (d) are applied at the output of the trimming control. The limits can be applied as limits to the controller output or as a reduced gain at the point the trimming control enters the basic system. If implemented by the "reduced gain" method, then manual control has a broader signal range, and the gain at the point of entry to the basic system becomes part of the overall loop gain. This allows an increased gain for the % O_2 controller.

The function generation for most installations can be a two-slope function generation as shown in Figure 16-7. On those installations where such a simple method is not sufficient, a more complex function generation based on multiple slopes or a polynomial equation is necessary.

For any trimming control, upon loss of flame or when operating at a boiler load below that of the minimum air flow limit, the controller has no way to control the excess combustion air. If these operating possibilities are not taken into consideration, then unsafe operating conditions can occur when the controller again becomes effective. The controller action in these situations should be carefully analyzed, and security action should be designed into the system. As a minimum, the controller should be automatically transferred to the manual or tracking mode. Additionally, the system may be designed to automatically revert the controller to the automatic mode when boiler operation is again within the range for controller operation and other requirements for automatic operation are satisfied.

16-6-2 Total Combustible or Carbon Monoxide in the PPM Range as an Individual Control Index

Based on the original theory of a constant ppm CO set point at all boiler loads, which has been previously described, the control arrangement is shown by the solid lines of Figure 16-8. The dotted lines show the addition to this control that would be necessary for operation at an optimum minimum fuel loss. The $f(x)$ function would be determined by factors of fuel analysis, excess air, flue gas temperature and actual boiler testing to determine the shape of the ppm CO/excess air curve.

In this control application, the control loop uses a simple feedback controller (a) with an adjustable external set point for ppm CO. The feedback measurement (b) of ppm CO or combustible gas usually has a measurement range of 0 to 1000 ppm. For the same reason described in the % oxygen control, limits (c) prevent the controller from causing unsafe boiler operating conditions upon a failure of the flue gas analyzing system. Also as in the % oxygen trimming control, the limits can be applied to the controller output or by use of "reduced gain" at the point the trimming control enters the basic control system.

When a control loop as shown is tuned for stable operation, the gain will probably be a very low value with the integral setting also at a low value. A typical relationship is that when the ppm CO is at set point a change in ppm CO of 100 ppm (10% of measurement range) is accompanied by a change in % oxygen of approximately 0.1 %. Since 0.1% O_2 is equivalent to approximately 0.5% total combustion air, then 100 ppm CO change will occur if the total combustion air is changed by approximately 0.5%.

Relating this to controller tuning, an approximate controller gain limit is 0.005/0.1 or 0.05. The values given should be considered as general guidelines only, since the equivalents may be as small as 75 ppm CO or less to 0.1% O_2 or as large as 150 ppm CO or more to 0.1% O_2, depending on boiler load, fuel being fired, and excess air level.

For the integral portion of the controller, the same time constants that exist in the % O_2 control loop must be considered for the ppm CO control loop. Of these, the "pure delay" time constant for transport of the flue gases from the combustion chamber to the measurement point may be slightly greater. This occurs if an "across the stack" analyzer is used, and it must be installed further downstream in the flue gas duct system. A difference in time constant of the measuring analyzer may also affect the relative settings of the intregral (repeats per minute) tuning.

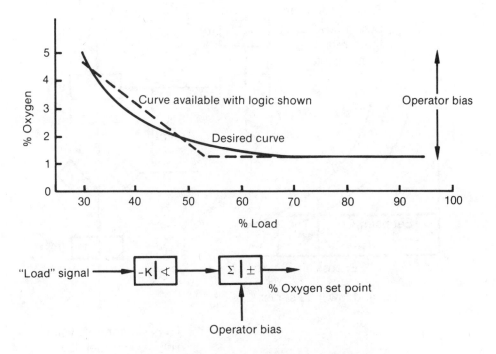

Figure 16-7 Capability of Elementary f(x) Logic

In addition to the security action of limiters (c), the same security actions for operation below the minimum air flow set point or upon flame failure should also be taken with this type of trimming control. As a minimum, the control loop should be reverted to the manual operating mode should either of these situations occur.

Other situations may occur that call for blocking control action from a ppm CO or ppm total combustible controller. If smoke is excessive, the controller must be inhibited from further reducing the excess air. When using ppm CO control, the presence of hydrocarbon combustible gas in excess of that of carbon monoxide should prevent any further reduction in excess air. For stoker-fired boilers, the increased furnace temperature as excess air is reduced may cause excessive grate temperatures.

16-6-3 Trimming Control Based on a Combination of % Oxygen and PPM CO

Early installations of ppm CO trim control (beginning in early 1970's) demonstrated some of weaknesses in this approach. In addition to the points above, CO or combustible gas formation may be affected by rapid changes in firing rate or excess air levels. Using the above control, poor burner performance can cause an excessive air flow change until enough combustion air has been added to dilute the flue gas back to the ppm CO set point. With the control set point at the minimum limit of % excess air, relatively small analyzer errors may drive the boiler operation into an excessive unburned fuel situation.

If boiler slagging or some other excess air limiting factor should be the limiting factor, then the ppm CO/% oxygen curve at the indicated set point is relatively flat. Such a relatively flat curve would greatly reduce control effectiveness.

Without other means of monitoring excess air level, improper operation may continue for an unlimited period. In addition, the potential benefit of a few percent less excess air resulting from basing the control on ppm CO is lost. To avoid inadvertent fuel losses from

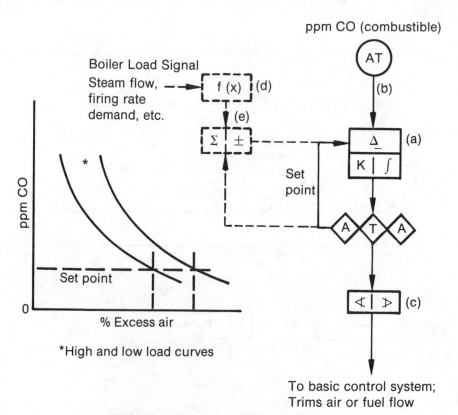

Figure 16-8 PPM CO Trim Control Loop

such occurrences, the level of excess air in the flue gas should be monitored by measuring % oxygen.

From measurement to combining % oxygen with ppm CO in a trimming control loop is a simple step. Two relatively simple control arrangements — of the many used or suggested — of flue gas analysis trimming control loops based on both % oxygen and ppm CO are shown in Figures 16-9A and 16-9B.

In operation, the control normally operates from ppm CO within a band of % O_2. At either a high % O_2 or low % O_2 limit, the control action switches to a % oxygen control. Since the desired excess air is a function of boiler firing rate, these % O_2 limits should be shifted as a function of firing rate. In this control arrangement, the difference between the high and low % O_2 limits is approximately 1% oxygen (the approximate band equivalent to 1000 ppm CO). The low limit should be at least 0.2% to 0.3% O_2 to allow the low % O_2 control loop to function.

The control arrangement shown in Figure 16-9A operates as follows. For any change in either of the two measured variables there will be a change in the other variable. The approach shown is to match the effects of % O_2 and ppm CO so that a single controller can be used. The boiler firing rate vs % O_2 relationship is developed by the function generation of items (a) and (b). This function is the low % O_2 limit and is determined by boiler testing. This function is balanced against the % O_2 analyzer (c) signal to produce an error signal in the difference logic (d). The width of the % O_2 band is determined by the positive bias (e), which produces the upper % O_2 limit. The value of bias (e) is determined by the range of the %

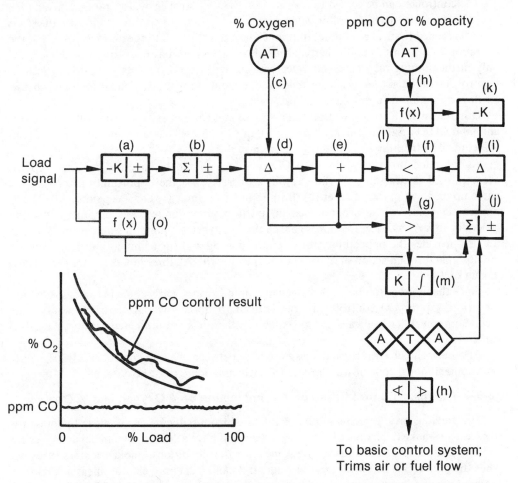

Figure 16-9A Trim Control, % Oxygen plus PPM CO or % Oxygen plus Opacity

oxygen signal and the desired width of the band between the high and low % oxygen set points. If the desired band is 2 percent oxygen and the % oxygen signal range is 0 to 5% at 1.0 gain on (c), the band would be 40 percent and the positive bias (e) would be 40 percent.

Changes in ppm CO are in the opposite direction from those of % O_2. The ppm CO signal from analyzer (h) is inverted and matched in range to the % O_2 signal in the negative proportional logic (k). An error signal of ppm CO relative to its set point is developed in the difference logic (i). Since ppm CO is highly nonlinear, the ppm CO signal is linearized in function generator (l). The desired ppm CO set point is determined by function generator (o) and potentially biased by the operator in summer (j). The result is a linear ppm CO error signal from the difference logic (i) to match the linear % O_2 error signal.

From this three potential error signals are produced, each of which are essentially linear and have approximately the same control response and gain characteristics. Bumpless automatic selection of the desired error signal is accomplished by the high and low selector logic, (f) and (g).

The selected error signal enters the controller (m), which then produces the trimming control signal. Using a single controller avoids any control windup problems when the control action is switched from % oxygen control to ppm CO control and vice versa. As in the other trimming control arrangements, the control action should be limited as shown in the high/low limits of limiter (n). Alternately, the limiting approach of "reduced gain" where the signal enters the basic control system can be used. Tuning of the controller is the same as that used with % O_2 control.

The controller can be forced to operate as a % O_2 controller by arbitrarily adjusting the ppm CO set point to a high or low value. Under steady-state operating conditions, the % O_2 control is tuned first. The set point of ppm CO is then adjusted to the desired value, and the operation as a ppm CO controller is checked. If the system has been properly aligned, the only further tuning may be a small decrease in the integral (repeats per minute) setting. Control should be automatically switched to the manual mode when the firing rate is below air flow minimum or upon flame failure.

Other arrangements of this trimming control make use of separate controllers for % O_2 and ppm CO. The arrangement shown in Figure 16-9B uses cascade control from the ppm CO control to trim the % O_2 control set point.

As indicated previously, the ppm CO set point should be load-programmed. This is accomplished in function generator (f). The operator has the opportunity of biasing the curve up or down in the summer (d). Both inputs to summer (d) have a gain of 1.0. The % oxygen set point is load-programmed in function generator (e). The output of function generator (e) acts as a feedforward signal to the % oxygen set point in summer (c). Summer (c) also provides the opportunity for operator biasing and for trimming the % oxygen set point signal through action of the ppm CO controller (b). All three inputs to summer (c) have a gain of 1.0.

Since the time constants for % oxygen and ppm CO are nearly equal in this cascade loop the primary ppm CO control must be detuned to obtain overall control stability. The secondary controller is always % oxygen control and has the tuning constraints described earlier.

Another control arrangement uses switching between the two controllers. In designing such control loops, control windup and the switching logic must be considered.

16-6-4 Trimming Control Based on a Combination of % Oxygen and % Opacity

The previous arrangements should be inhibited from reducing excess air in the presence of excessive smoke (high opacity). All fuels except gas have a smoke potential in an excess air condition. In some installations and/or some operating conditions, smoke becomes excessive while the ppm CO is below the desired value. If it can be determined that this is always true for a particular installation, then % opacity can substitute for ppm CO in the control arrangement in Figure 16-9A and 16-9B. This is based on the marked similarity of the curves of ppm CO vs excess air and % opacity vs excess air shown in Figure 16-5.

This application has been used particularly for spreader stoker applications. In these applications the low % O_2 limit is usually that required to prevent excessive furnace or grate temperatures. The high % O_2 limit is that which produces a minimum opacity level. Additional excess air may cause particulate carryover due to flue gas velocity. Except for these variations, the calibration, tuning, and alignment procedure of this control arrangement is the same as that using ppm CO as the feedback signal.

16-7 Limiting Factors in Reducing Excess Air

The preceding parts of this section discuss the use of flue gas analysis control as a means of improving boiler efficiency. The boiler efficiency is not always the issue, and the flue gas analysis is not always the limiting factor. It is possible to improve boiler efficiency while causing the performance of the overall system, which includes the boiler, to deteriorate.

If a boiler is used to furnish steam to a power generation turbine, the gain in boiler efficiency may be at the expense of turbine performance. It is well known that such installations use superheated steam in practically all fossil fuel installations. It is also well known that reducing the excess air of a boiler may cause steam temperature to be reduced unless some steam temperature control mechanism is in the active control range (refer to Section 7). A reduction in steam temperature reduces the thermodynamically available energy for con-

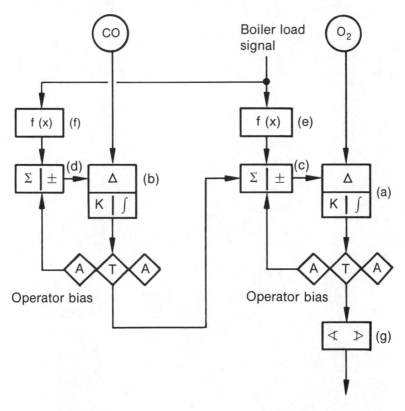

Figure 16-9B Trim Control % Oxygen plus PPM CO

version to power. It has been shown that up to the point of excessive combustible gas formation, if excess air reduction reduces the steam temperature, then the overall result is a decrease in boiler-turbine system performance.

Reduction in excess air also results in increased furnace temperature. For coal-fired boilers, this can increase furnace slagging and clinker formation (fusing of the ash). The result is difficulty in ash removal. Another problem related to furnace temperature is that of excessive grate temperature of stoker-fired boilers. An excess air level above that for minimum fuel consumption may be necessary for installations with this problem.

The percentage of carbon in the refuse is another factor to be considered when the boiler is fired by coal, wood, or other solid fuel. Representatives of boiler manufacturers have stated that they can find no specific relationship between the level of combustible gas in the flue gas and the % carbon in the refuse.

The consequence is that there are several limiting factors to the reduction in the level of excess combustion air. The economic level of the combined effect of excess air and combustible gas is only one of these limiting factors. All such factors should be carefully considered before a decision to apply control equipment based on analysis of the flue gases. The control application method and its implementation should be based on knowledge of the particular limiting factors that may apply.

Section 17
Combustion Control for Liquid and Gaseous Fuel Boilers

The basic difference in the approach to combustion control for liquid or gaseous fuel boilers as opposed to that for solid fuel boilers is that the the fuel can easily be measured. This basic difference applies, however, only for systems that incorporate fuel flow/air flow ratio or difference as part of the control strategy. Simple systems such as the single-point positioning (jackshaft) or parallel positioning systems can be applied to all types of boilers in a similar fashion.

As discussed in Section 7, the combustion control loops for all boilers respond to the Btu demand signals generated in the master control loop. The Btu demand signal is assumed to be linear with respect to Btu flow. A fully modulating control is used for almost all industrial applications.

17-1 Single-Point Positioning Control

As shown in Figure 17-1 a single-point positioning or "jackshaft" is a mechanical system. The position of the fuel control valve and the combustion air flow damper are connected together in a fixed relationship and move in unison to the demands of the master regulator.

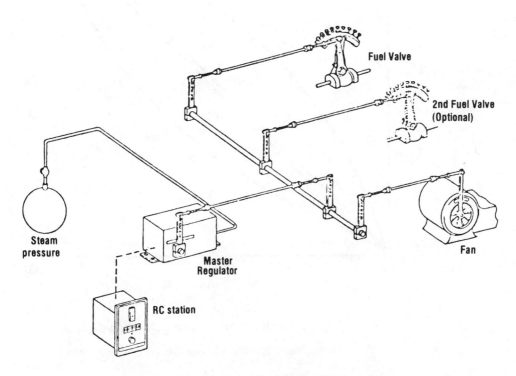

Figure 17-1 Single-Point Positioning Control System

A basic requirement of this type of system is the careful mechanical alignment of the fuel valve and the air damper positions. Fuel valves and air dampers tend to have different flow characteristics. Typical characteristics are shown in Figure 17-2. If the master regulator were to move each to the 50% position, then air flow for approximately 75% capacity would be provided while fuel for 25% capacity was being supplied. By making the flow characteristics linear, they can then be aligned.

In the case of the air damper, the alignment tool is the use of linkage angularity as discussed in Section 11. In the case of the fuel flow, the control valve is usually supplied with a cam arrangement for changing the perceived flow characteristic. The procedure is to linearize the air flow characteristic and then to match the fuel flow characteristic to that of the air flow. To perform this alignment procedure properly, it is necessary to perform combustion tests at several different boiler loads.

On the surface it appears that such a proper alignment would complete the requirements and that no further improvement is necessary. One weakness of this system from a fuel/air ratio standpoint is that the position of the fuel valve is not always a true measure of fuel Btu flow. Another weakness is that the fan damper or inlet vane position is not always a true measurement of the flow of oxygen for combustion.

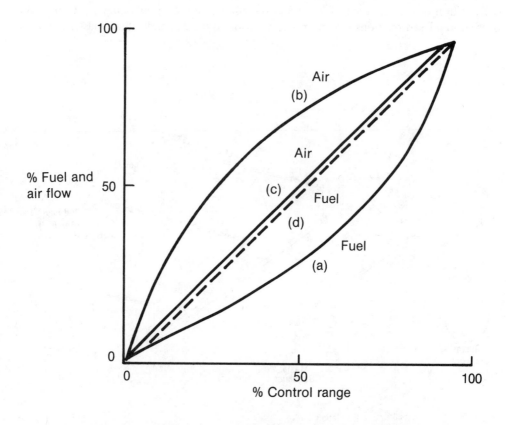

a, b — Basic flow characteristics of controlled devices

c, d — Characteristics after linearization and alignment

Figure 17-2 Flow Characteristics of Valves and Dampers

A variation in either the flow of oxygen or fuel changes combustion conditions and the ratio of excess air for combustion. With the parallel arrangement discussed and using gaseous fuel as an example, this can happen in several ways:

(1) Change in the fuel unit (scfh) Btu value. The fuel flow in scfh may be constant but total Btu flow changes.

2) Change in fuel temperature. The fuel density changes and the fuel flow volume (scfh) changes. Total Btu flow changes even though unit Btu value remains constant.

(3) Increase or decrease in fuel specific gravity. The fuel density changes and simultaneously the unit Btu value changes. These combine to change total Btu flow.

(4) Increase or decrease in fuel pressure. This causes density to change, thus changing fuel flow volume (scfh). It also changes scfh fuel volume through the changed pressure drop across the control valve. These combine to change total Btu flow.

(5) Increase or decrease in combustion air temperature. This changes the air density and delivers a changed amount of oxygen to the combustion process. Change in the density also affects fan delivery pressure and the total flow of combustion air.

(6) Change in the humidity of the combustion air. This changes the percentage of dry air in the total air, thus changing the flow of oxygen to the combustion process. The density of the air flow is also changed, affecting the fan delivery pressure and differential across the flow damper, thus affecting air flow.

(7) Changes in atmospheric pressure. This changes the fan total air flow delivery and pressure, thus affecting oxygen flow to the combustion process.

Since the fuel control valve and the combustion air damper are mechanically linked and the system does not include measurement of any of the above variables, the base system as shown cannot compensate for these variations. If we wish to compensate for these variations in fuel/air ratio, it is necessary to modify either the fuel pressure, the fuel control valve position, the combustion air control damper position, or the fan speed.

With the basic systems above, the fuel/air ratio may vary over a control error band of up to approximately 40% excess air. If the control system is adjusted for too low an excess air level, the control error band may at some time cause the boiler to operate with insufficient excess combustion air to burn all the fuel. Under such circumstances fuel may be wasted at 5 to 6 times that which would occur if the excess combustion air were too great by the same amount. With such systems it is, therefore, good practice to calibrate the system with sufficient excess air to accommodate the control error band.

Though the single-point positioning system is mechanically linked, a flue gas analysis trimming control loop can be applied to control the fuel/air ratio and reduce the control error band. Figure 17-3 shows some typical arrangements for applying the trimming control. Other methods include changing the length of the link to the combustion air flow damper, changing the length of the drive arm connected to the link, or both.

In all of these arrangements the control signal from the trimming control originates in the control loops covered in Section 16. If there is a significant error in the basic system, the trimming control requires time to make the adjustment. If the basic system error is different at one load than it is at another load, time is required to readjust as the load is changed. This is usually not a problem unless the boiler load changes rapidly. In such cases the trimming control may have difficulty keeping up with the changes in excess air. This problem is demonstrated in Figure 17-4.

As shown in this plot, at 50% load, the trimming correction due to error in the basic system is 25% and is 12.5% at 25% boiler load. The controller output, therefore, must change by 12.5% as the boiler load is changed. This type of control is relatively slow due to the time delays described in Section 16. Some microprocessor based "trim" control arrangements use memory of last time at this load vs. trim signal relationships to help move quickly to a new output signal as the load is changed. The plot also demonstrates that poor alignment of the

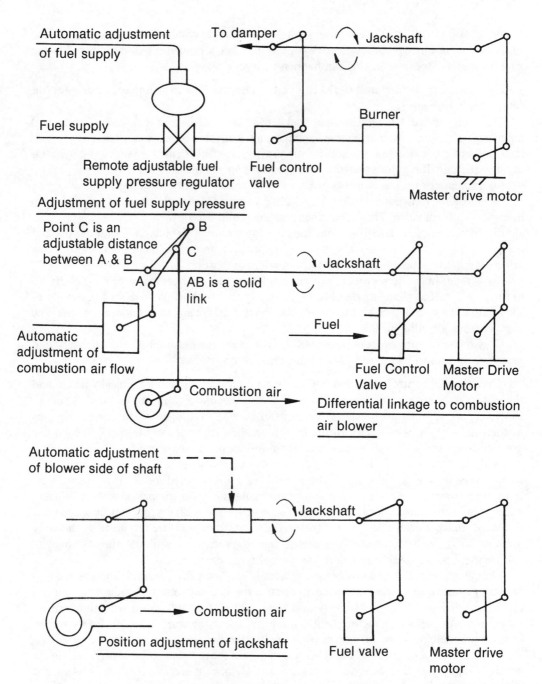

**Figure 17-3 Automatic Control Methods for Changing
Fuel-Air Ratio in Mechanical Control Systems**

basic system may create the need for an excessive amount of correction from the flue gas analysis trim control. In such cases trim control limits may prevent the amount of correction that is needed.

17-2 Parallel Positioning Control

The functions shown in the mechanical single-point positioning arrangement can be performed using instrumentation control. Such systems are called parallel positioning control systems and link the functions pneumatically or electrically. A SAMA control logic diagram of a parallel positioning system comparable to the single-point positioning system of Figure 17-1 is shown in Figure 17-5.

Such control systems must be aligned in the same manner as the single-point system. In this case, a cam in the positioner of the fuel control valve is used to linearize fuel flow and align it with air flow. For air flow calibration, linkage angularity and/or the cam in the positioner of the damper operator may be used to linearize the air flow signal vs. flow.

A parallel positioning system has the same weaknesses and same control error band as the single-point system. Note that such a system may have a simple means of biasing the fuel/air ratio through use of the manual loading function, (a) in Figure 17-5. This adjustment means is useless without the use of a flue gas analyzer or some other form of combustion guide. In order that the system be aligned with the operator adjustment in the midpoint of its range, the firing rate demand input to summer (b) is set at a gain of 1.0. Assuming that the operator is provided a plus or minus 15 percent bias, the gain of the input from (a) is set at 0.3. With both inputs at 50%, and with no bias, the output of summer (b) would be 65%. The

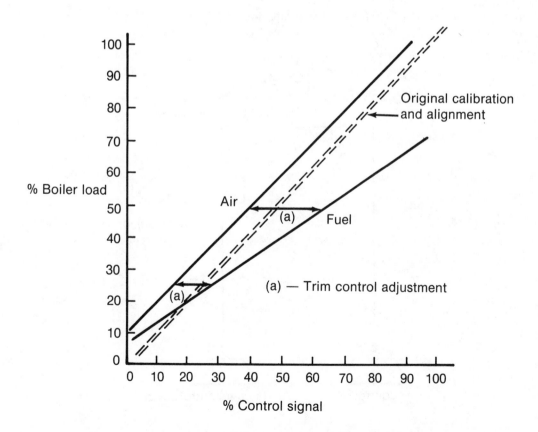

Figure 17-4 Alignment Problem: Air Is Cold, Fuel Pressure Is Low

bias in summer (b) is then set at minus 15% so that the signal to the air damper control device will always match the firing rate demand signal.

The ease with which a boiler operator makes such a fuel/air ratio adjustment also makes it easier for the operator to cause system misalignment. One advantage of the parallel system is that the timing of a fuel flow change or an air flow change can be modified by inserting a time constant into either of the two control signals to improve matching of actual fuel and air flow to the furnace. This makes possible improved dynamic operation.

Improvement of the control system in order to narrow the control error band is accomplished by the use of flue gas analysis trimming control. Since the basic system is an instrument control system, connection of the trimming control to the parallel positioning system is usually simpler than when connecting to a single-point positioning system. The arrangement of a parallel positioning system plus trim control is shown in Figure 17-6.

In the arrangement in Figure 17-6, the control signal (a) to fuel is used as the load signal in the flue gas analysis trimming control. The output of the trimming control (b) modifies the basic fuel control signal in the multiplier (c). The proportional plus positive bias (e) reduces the gain of the trimming control signal (b) and positions the output from (e).

Assuming that the trimming effect is to be an air flow multiplication of 0.85 to 1.15, the gain setting of the proportional plus bias (e) would be 0.3 and the bias would be 0.85. This would provide a multiplication of 1.0 at the midpoint of the flue gas analysis trim control output. The output of the multiplier (c) is a modified basic signal that acts as the control signal for air flow. If the furnace is a balanced draft furnace, the connection to the furnace

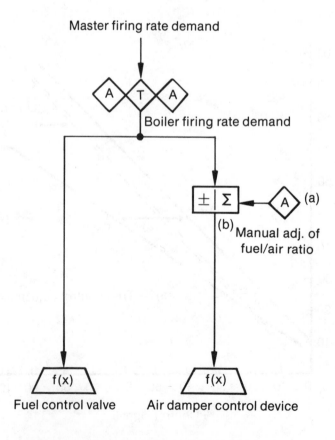

Figure 17-5 Parallel Positioning Control System

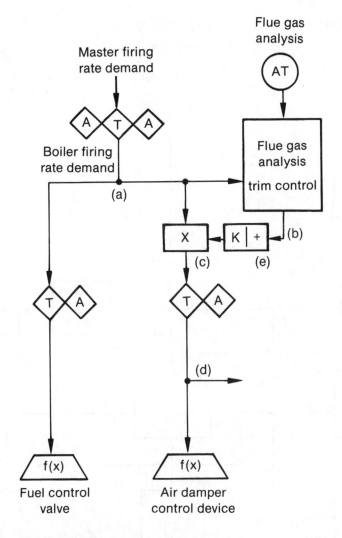

Figure 17-6 Parallel Positioning Control with Flue Gas in Analysis Trim

draft control loop is shown at (d). This connection is not necessary if the furnace draft is controlled with a simple feedback control loop.

In some cases a summer is substituted for the multiplier (c). This is theoretically incorrect since the effect of the flue gas analysis trim control would then be greater at lower boiler loads than at higher boiler loads.

17-3 Metering Control Systems

The weaknesses of the basic single-point positioning and parallel positioning systems can be overcome by including measurements of fuel and air flow in the control strategy. The evolution of several application methods for such metering systems has resulted in what is now generally recognized as a standard control arrangement. This control arrangement shown in the Figure 17-7 block diagram also includes active safety constraints. Such an arrangement is suitable for any liquid or gaseous fuel or fuel combination in which the unit Btu values do not vary by significant amounts (more than approximately 10%). Several names, which all designate the same control logic, have been ascribed to this system. Such names are, "cross-limited", "lead-lag", "self-linearizing", and "flow-tieback".

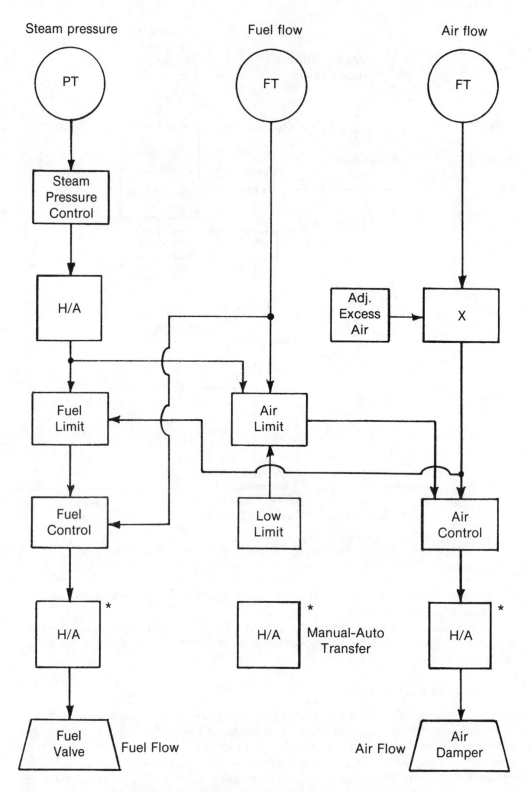

Figure 17-7 Metered Cross-Limited Boiler Control System

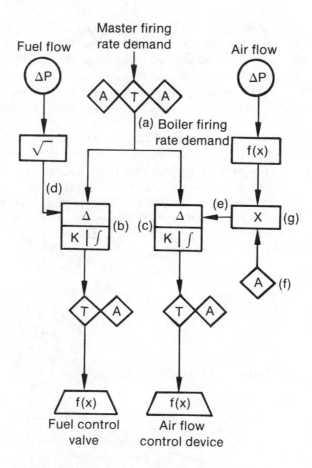

Figure 17-8 Basic Metering Boiler Control System

In the type of system shown in the SAMA diagram of Figure 17-8, the firing rate demand signal (a) acts as a common set point for the fuel flow controller (b) and the air flow controller (c). Since the fuel flow measurement signal (d) and the relative air flow measurement signal (e) are linear, the base fuel/air ratio is established by the calibration of the air flow measuring device. As described earlier the relative air flow measurement is calibrated by combustion tests at various boiler loads. The result is that when the combustion conditions are correct, the percentage reading of both the fuel flow and the air flow measurements are equal. The manual signal (f) and the multiplier (g) provide a convenient means for the operator to alter the calibration of the air flow measurement and thus modify the fuel/air ratio.

To the basic parallel flow controller arrangement described above, high select, low select, bias, and gain functions are added as shown in Figure 17-9. These functions add active safety constraints to the system. The low select function (h) compares the firing rate demand signal (a) to the air flow measurement signal (e) and the lower of the two becomes the set point of the fuel controller (b). The result is that the fuel flow set point is limited to the level of the signal representing available combustion air flow. Similarly the high select function (i) forces the air flow set point to the higher of the two signals that represent firing rate demand and fuel flow. The result is that actual fuel flow sets the minimum air flow demand.

The bias and gain functions (j) are added to provide a small 3 to 4% dead band between the application of the high and low select functions. This addition prevents the effect of process measurement "noise" from causing interaction between the fuel flow and the air flow control loops. The addition of the manual signal (k) connected to the high select (i) provides a minimum air flow control capability by preventing the air flow set point from being reduced below 25% of full range. The 25% minimum air flow setting is one part of the NFPA standards.

This diagram also shows the application of a flue gas analysis trim control loop replacing the manual fuel/air ratio adjustment. By its control action, the air flow measurement signal (e) is continuously calibrated so that the programmed flue gas analysis will be obtained. The firing rate demand signal (a) is used as the "boiler load" signal in the arrangement shown. The fuel flow signal (d) or a steam flow measurement signal could also have been used. The points marked A, B, and C are the points at which this control system arrangement can easily be modified for application to the use of other fuels or fuel combinations or for other air flow control arrangements as covered in Section 13.

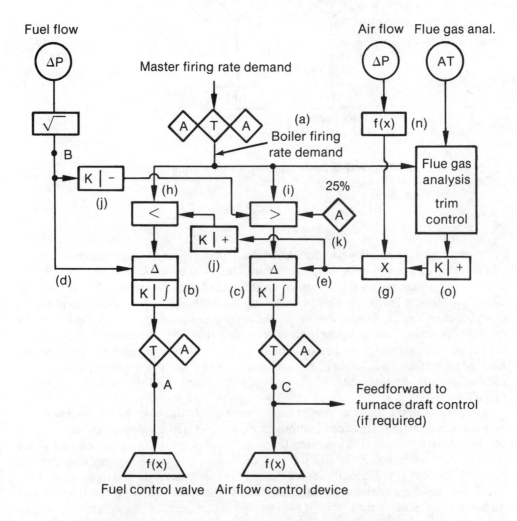

Figure 17-9 Metered Cross-Limited Combustion Control System

The control system alignment consists of:

(1) Calibration of the air flow function generator (n).
(2) Setting the bias values of approximately plus and minus 25% and gains of 1.0 to items (j).
(3) Setting the minimum air flow signal (k) at 25%.
(4) Setting the gain of proportional plus bias (o) to 0.3 and setting the bias to plus 85%.

After the alignment procedure, the next activity is to tune the flow controllers (b) and (c). This procedure is typical of the tuning of any flow control loop. The optimum gain setting will probably be in the 0.5 range and the optimum integral setting will probably be several rpm (repeats per minute). Dynamic load change testing may reveal a deviation in the desired excess air during the load change. In this case, either the fuel flow or air flow control loops should be detuned until their dynamic response is the same.

The 25% bias values of items (j) avoid any effect from the limiting control during this tuning period. After the flow controller tuning operation these settings are adjusted to their operational settings. This is accomplished by gradually reducing the bias values until interaction with the flow controllers is indicated. This will normally show up best as the process noise band of flue gas analysis.

In some installations it may be necessary to alter the arrangement shown in Figure 17-9 because of excessive process flow measurement noise or the need for very rapid firing rate changes. By revising the control application to that shown in Figure 17-10, the primary control response is that of a feedforward system with a reduced-gain feedback trimming control from the flow control loops.

The additional items in the control logic are the summer and bias units (m) and (l). Both inputs to each of the summers are set at gains of 1.0. This allows the detuning of the flow control loops to accommodate the process flow measurement noise without affecting the response of the system. Since the feedback portion of the control must provide only a small portion of any control output change, the flow controllers may be considerably detuned while still obtaining stable and responsive control.

One additional requirement has been added. In the system shown in Figure 17-9 it was not necessary to parallel the control signals to the fuel control valve and the air flow control device. In the system shown in Figure 17-10 the initial feedforward signal attempts to position these devices to obtain the desired fuel/air ratio. It is, therefore, necessary to calibrate these devices for matching flow vs. control signal characteristics.

If a single fuel such as fuel oil is used instead of fuel gas, the modification is that of exchanging a linear fuel gas measurement (d) for a linear oil flow measurement and substituting a fuel oil control valve for the fuel gas control valve. Figure 17-11 shows the functions connected to points A and B when using return-type fuel oil burners. In this case, the "supply" and "return" flow measurement transmitters must be very carefully calibrated so that the measurement errors in the individual flows do not accumulate in the signal representing oil flow to the furnace.

Other potential modifications at points A and B are those for combination fuel firing. An example of this is the combination of fuel oil and natural gas or the combination of natural gas and a relatively stable process-generated gas such as coke oven gas. The modification for burning either fuel oil or natural gas is shown in Figure 17-12.

In this modification the two fuels are totalized in the summer (o) on a basis that combines the effect of difference in air required for combustion and that of differences in excess air requirement. The gain value applied to the individual fuel flow signals into the summer (o) are calculated as follows. Assume that there are two fuels, natural gas @ 1,000 Btu/scfh and fuel oil @ 19,000 Btu/lb. The natural gas flow range is 0 to 200,000 scfh and the fuel oil flow range is 0 to 10,000 lb/hr. The primary fuel is used as the base input with a gain of 1.0.

Assuming that the primary fuel is natural gas, the 0 to 100% range of the total fuel flow signal is therefore 0 to 200 MBtu/hr. The oil flow range is 0 to 190 MBtu/hr. On a theoretical basis fuel oil requires approximately 7.3 lbs of combustion air per 10,000 Btu, while natural gas requires approximately 7.2 lbs of air per 10,000 Btu. Assume that the base of total combustion air for the natural gas is 110% (10% excess air) and 115% total air (15% excess air) for the fuel oil. The gain for the fuel oil flow input to the summer (o) can then be calculated by the formula

Gain =(190/200)*(7.3/7.2)*(115/110).

The result is 1.0065. If there were 3 or more fuels, the primary fuel would be the base and the other two or more would be individually matched to the primary fuel.

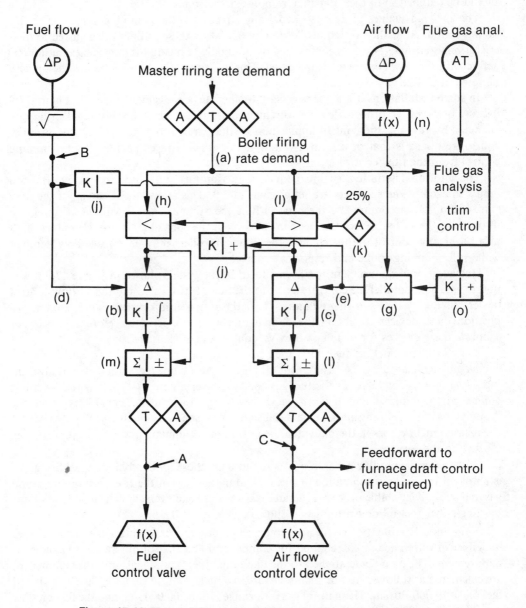

Figure 17-10 Metered Cross-Limited Combustion Control System

The totalized fuel flow signal (p) then enters the control system in the same manner as the flow measurement of a single fuel. The interlock shown (n) is to prevent the fuel oil flow signal (q) from being used when fuel oil is being circulated prior to burner light off.

This control arrangement should properly be used only when there is a capability for two fuels, but they are burned one at a time. Switchover from one fuel to the other can be accomplished properly on automatic control if only one fuel is on automatic. To use this arrangement with simultaneous automatic firing of both fuels causes the fuel flow control loop gain to double due to doubling the fuel capacity available to the total fuel control signal.

Simultaneous automatic firing of two or more fuels without altering the fuel control loop gain can be accomplished with the point A and B modifications shown in Figure 17-13. In this arrangement the fuel control signal is split so that the capacity available to this signal does not change. The manual signal (r) sets a ratio for one of the fuel control signals relative to the total fuel control signal. The delta block (s) subtracts this signal from 100% of the signal value. The result is that the sum of the two fuel control signals is always equal to the fuel control signal from controller (b) of the system arrangement shown on Figure 17-9 or from summer (m) of the system arrangement shown on Figure 17-10.

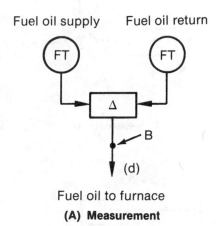

(A) Measurement

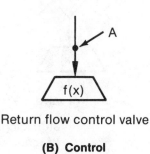

Return flow control valve

(B) Control

Figure 17-11 Modification to Basic Combustion Control System for Fuel Oil Firing in Return-Type Burners

An alternate control modification that will hold the fuel control loop gain constant while the boiler is simultaneously firing a combination of fuels is shown Figure 17-14. In the arrangement, the fuels are totalized on an "air required" basis as in Figure 17-12 and 17-13. The control signals to the individual valves are added in the summer (t) and balanced against the basic fuel control signal in a controller (u), which produces the control signal to the control valves.

The fuel control valves will probably be of different capacities, the fuel stream of different fuels will have different Btu values, and the number of fuels being used at any one time must be accommodated. This is accomplished by adjusting the input gains of the individual valve control signals into the summer (t). These gains are the ratio of the individual valve capacities in Btu value to 100% of the desired total Btu range. The desired total Btu range is that

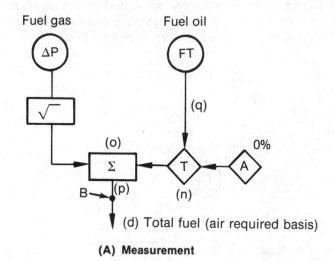

(A) Measurement

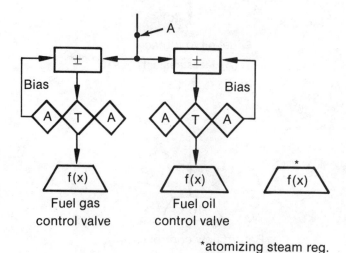

(B) Control

Figure 17-12 Modification to Basic Combustion Control System for Fuel Gas or Fuel Oil Firing (Not in Combination on Auto)

required for achieving full boiler load plus necessary overfiring capability. For example, if the base fuel had 100% capability and two auxiliary fuels each had 50% capability, the gain for the base fuel would be 1.0 and each of the others 0.5.

When waste process-generated gases are available, it is usually desirable to burn these gases on a priority basis before using purchased fuels. Figure 17-15 is an example of one method of "as available" or "priority" fuel control, shown as a modification at points A and B of the basic control arrangement in Figure 17-9.

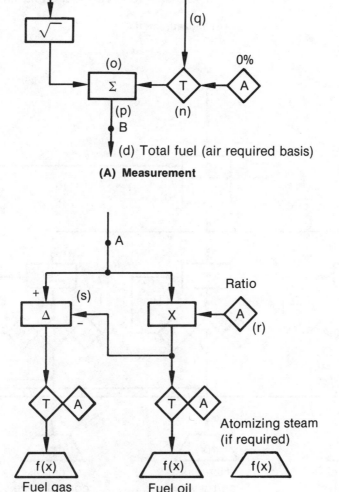

Figure 17-13 Modification to Basic Combustion Control System
for Firing Fuels in Combination

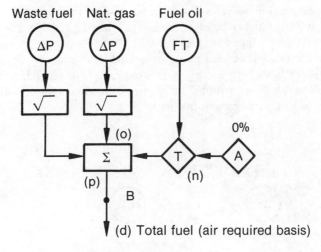

(A) Measurement

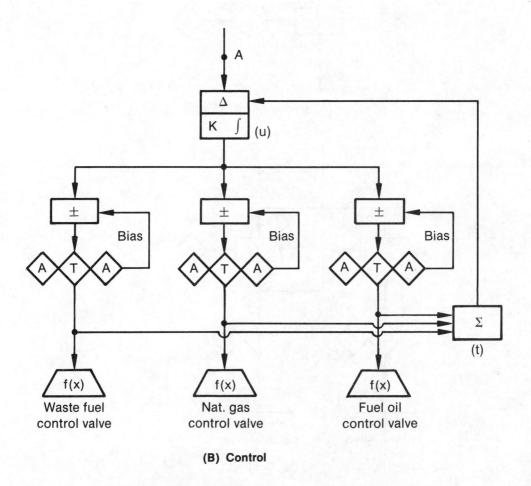

(B) Control

Figure 17-14 Modification to Basic Combustion Control System for Combination Fuel Firing

The pressure controller (v) is applied to the waste gas system. In operation with sufficient waste gas available, the pressure of the waste gas is at or above the set point and the output of the pressure controller is 0. At this time the signal to the waste gas control valve subtracts from the signals to the fuel gas and fuel oil control valves, placing these signals at a minimum value with only a pilot flame sustaining combustion of these fuels. With less waste fuel capability than that required for the boiler load, the pressure of the waste gas will fall below the set point of controller (v) causing its output to increase. This subtracts causing the output of delta (w) to decrease and reduce the waste gas flow until the pressure controller (v) is satisfied.

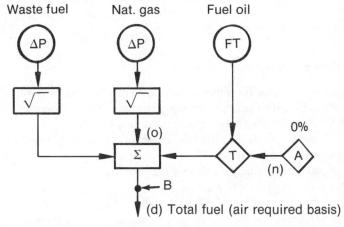

(A) Measurement

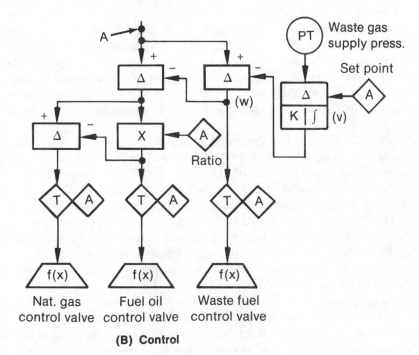

(B) Control

Figure 17-15 Modification to Basic Combustion Control System for "Priority" or "As Available" Waste Fuel

The difference between the input from A and the output of delta (w), which is equal to the output of the controller (v), is then added to the fuel gas and fuel oil control signals. This satisfies the total firing rate demand with the waste gas as a priority fuel.

In the previous arrangements involving combination fuels, it is assumed that it is not necessary to ratio the flows of the fuels involved. If the ratioing of flow is needed for "least fuel cost" control action or for other reasons, then the basic control of Figure 17-9 is altered in accordance with Figure 17-16. The flows are totalized as before on a "total air required" basis. The output of this totalization acts, however, only on the active constraint of "high select" function, which forces the air flow set point to the value of that required for the total fuel being burned. For the normal control functions, the fuel flow feedbacks respond to the individual set points of their controllers. The set points are generated through a set point ratio arrangement that is the same as that used for splitting the control signal as shown in Figure 17-13. The ratio set may be generated manually, as shown, or through other control functions that are not shown here.

17-4 Effects of Fuel Btu Variation

The control system arrangements shown above are not the most desirable arrangements if the fuel Btu varies significantly (more than approximately 10%). Assuming a gaseous fuel, a fuel unit Btu variation of 10% with the same volume flow will change the total Btu delivery to the furnace by 10% while the air flow remains constant. This causes a shift of 10% in total air for combustion. Such a deviation would normally be unacceptable, but is still within the range for a plus or minus 15% flue gas analysis trim control.

In fact, for hydrocarbon fuel gases a fuel Btu variation is accompanied by a specific gravity change. For the same volumetric flow, this alters the differential pressure across an orifice and tends to counteract the fuel-air ratio control effect of the unit Btu value change. If a hydrocarbon fuel such as natural gas is the fuel used, the net effect on % excess air for a 10% change in fuel Btu value is approximately 3.5%, the approximate threshhold of acceptability without flue gas analysis trim control.

If the change in Btu value were 20% or 30% as is often the case with the use of refinery gas, the change in fuel Btu would significantly alter the total Btu input to the boiler, changing the steam pressure and causing the firing rate demand signal to change. In correcting the total Btu input to its previous value, the combustion air flow would be adjusted to an incorrect value. A flue gas analysis trim control could be used to recorrect the air flow, but this would be after the fact of the process disturbance and the effect on excess combustion air.

A better solution is to alter the fuel flow before the process is disturbed. A flue gas analysis trim control that modifies the fuel flow or the fuel flow demand signal will prevent most of the process disturbance described.

Since the air flow required is nearly constant for a given total Btu flow, then a change in fuel unit Btu value (and thus total Btu input) will cause an immediate change in the flue gas analysis. The change in the flue gas analysis can therefore be used to control a multiplication that recalibrates the fuel flow or fuel demand signal so that total Btu value will be the same as before. While there is some dead time in this action, the flue gas analysis response is many times faster than the total boiler response. In this way the disturbance to the process is minimized. Such an arrangement is shown in Figure 17-17.

The system shown is the same as that of Figure 17-9 except for the application of the flue gas analysis trim. As shown, the flue gas analysis trim control loop drives a multiplication (x) of the fuel flow measurement signal. The multiplication range is a function of both Btu value and the offsetting effect of change in specific gravity on the the flow measurement. This arrangement is satisfactory if the maximum multiplication expected is not more than approximately 1.15 or less than approximately 0.85.

If the maximum multiplication exceeds the approximate 0.85 to 1.15 range, the change in the feedback loop gain becomes more significant. Expected multiplication ranges of greater than the approximate 0.85 to 1.15 range require the multiplication on the demand or set point side of the controller in order that the feedback loop gain will be unaffected. This change is shown in Figure 17-18. This diagram also shows a divider (z), driven by the flue gas analysis trim control. This divider, by keeping the total "air flow required" range of the fuel measurement constant, assures that the air flow limiting function will remain correct. The technique described above is also suitable for installations that burn combination fuels with one or more of the fuels having a variable Btu content.

The desirability of any variable Btu fuel technique is related to the amount of process interaction that is involved. The requirement is that the correction or compensation be handled within the control system without involving process temperature or pressure changes

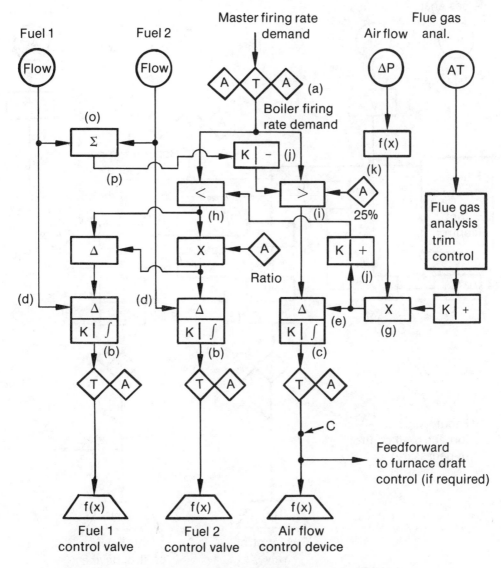

Figure 17-16 Metered Cross-Limited Combustion Control System with Adjustable Ratio Firing of Fuels in Combination

in order to obtain the necessary correction. The technique described and shown in Figure 17-17 and Figure 17-18 results in some process interaction since the process feels the Btu change and sends "after the fact" information on % oxygen back to the control system. The interaction period is based on the analysis and control time period, a fraction of the total process time constant.

An emerging technique that appears to be noninteractive with the process uses two fuel measurements in series. One measurement is an orifice type meter that is affected by specific gravity. The other meter is a volumetric meter such as a vortex shedding meter that is not affected by specific gravity. The result of a continuous calculation using a formula involving these two flows, the specific gravity-Btu value relationship, and the relationship to combustion air requirements can provide an immediate and proper total Btu feedback to the fuel flow controller.

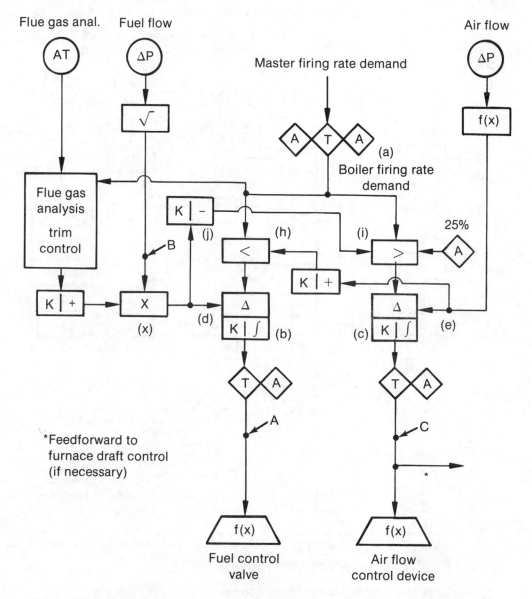

Figure 17-17 Metered Cross-Limited Combustion Control System — Variable Btu Fuel

Other techniques for controlling the combustion of variable Btu fuel are more complex than that shown in Figures 17-17 and 17-18. Among these are correction of fuel flow to mass flow (including specific gravity correction) and unit Btu measurement and compensation of fuel flow values. The measurements required have some time constants and their compensations should be fast enough so that they may correct the fuel flow before any appreciable disturbance to the process. When these variables cannot be measured, a slower method of using the boiler as a calorimeter may be used. This method is suitable for longer term fuel Btu variations and will be discussed in the section covering control systems for coal-fired boilers.

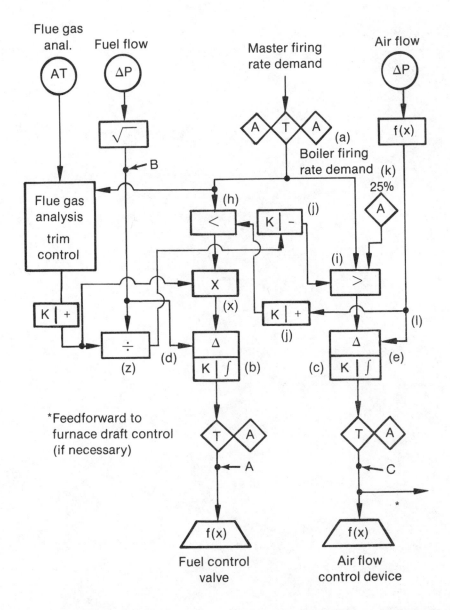

Figure 17-18 Metered Cross-Limited Combustion Control System — Variable Btu Fuel

Section 18
Coal or Solid Fuel Stokers

Coal or industrial solid waste is generally burned in boilers as a fuel bed on a grate. The mechanism is a mechanical stoker. Coal may also be finely ground and blown into the boiler furnace. The control characteristics of solid fuel stokers are quite different from those for pulverized coal. All types of stokers use a fuel bed or grate, but the control characteristics for different types of stokers are somewhat different. A basic difference that is related to the type of stoker is the amount of residence time for the fuel in the furnace prior to its combustion.

18-1 Solid Fuel Preparation for Firing

The various steps in the coal handling process are shown in Figure 18-1. While there are "mine mouth" power plants that burn the coal essentially as it is mined, coal used in industrial boilers generally follows the steps shown. The coal is cleaned and sized in the process shown. For proper burning of coal with a stoker the size of the coal lumps is important. The coal bunkers admit coal by gravity directly to the stoker hoppers.

The largest volume of non-coal solid fuel is wood waste or bark from the paper or lumber industry. The steps in its preparation and feed to the furnace are shown in Figure 18-2. The "hogger" device chops the wood waste to a somewhat uniform size so that it can be easily handled and burned by the stoker. Other solid waste fuel is bagasse, the fibrous residue from sugar cane, coffee grounds (which remain after the process of making instant coffee), and other solid refuse. The handling of these fuels is similar to that of wood waste. A characteristic of industrial solid waste is its high moisture content, at times in the 50 to 60% range.

18-2 Types and Classification of Stokers

Stokers are of three general types:

(1) Spreader stoker
(2) Underfeed stoker
(3) Overfeed stoker

In the spreader stoker, coal is flipped by a distributor at the bottom of the stoker hopper onto the grate. A portion of the coal is burned in suspension, with the heavier pieces falling to the grate.

The coal feed is a volumetric feed and is regulated by a control lever that adjusts the amount of coal admitted to the spreader. Combustion air is admitted through the fuel bed on the grate from underneath and is adjusted by a single control device. In order to increase the turbulence and complete the combustion, secondary combustion air is added as jets of overfire air above the grate.

There are three basic subtypes of spreader stokers. The fuel feed is the same for these: the difference is in the handling of the fuel bed and ash discharge, as follows:

(1) Dump grate — The grate sections are periodically turned at 90 degrees and the accumulated ash is dumped to the ashpit. A longitudinal section of grate is dumped at any one time. These stokers are usually built of three or more of these longitudinal sections.

(2) Travelling grate — In this type, shown in Figure 18-3, the grate moves slowly forward, discharging the ash to an ashpit at the lower front of the boiler.

(3) Vibrating grate (similar variations are the oscillating grate and reciprocating grate) — The grate is vibrated, causing the burning coal particles to move forward on the grate. The ash is discharged to an ashpit at the lower front of the boiler.

The heat flow response of the spreader-type stoker is quite rapid since a large percentage of the coal is burned in suspension. Some coal energy is stored on the grate and released gradually during the grate burning process. A wide variety of coals may be burned successfully with this type of stoker.

Because of the rapid response, a spreader stoker is useful for following fluctuating loads. Because of the suspension burning, the fly ash carryover is high, and low load smoke is a problem. This may require considerable attention to the overfire air and the reinjection of the

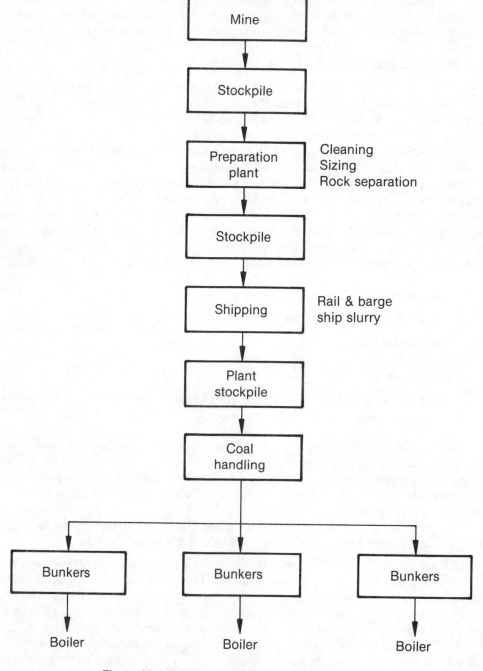

Figure 18-1 Fuel Preparation and Handling — Coal

fly ash and cinders to the grate. The spreader-type stoker is also the primary fuel-burning mechanism for burning wood waste or bark.

The second basic type is the underfeed stoker. For a small boiler, an underfeed stoker may be a "single-retort" type as shown in Figure 18-4. In this type a variable speed ram pushes new coal into the furnace from underneath the center of the retort. The burning surface of the coal is constantly being broken up by the new coal coming in.

Primary combustion air is fed to the burning zone through air tuyeres on each side of the center. Secondary air as overfire air jets above the burning zone is also necessary. The ash is discharged at the sides.

Maintenance is relatively high on this type of stoker. It will burn a variety of coals since the burning surface is constantly broken up. Because there is very little suspension burning, fly ash and cinder carryover are low. Since all of the fuel is stored in the furnace for a period before actual burning, most of the immediate response is due to a change in combustion air flow.

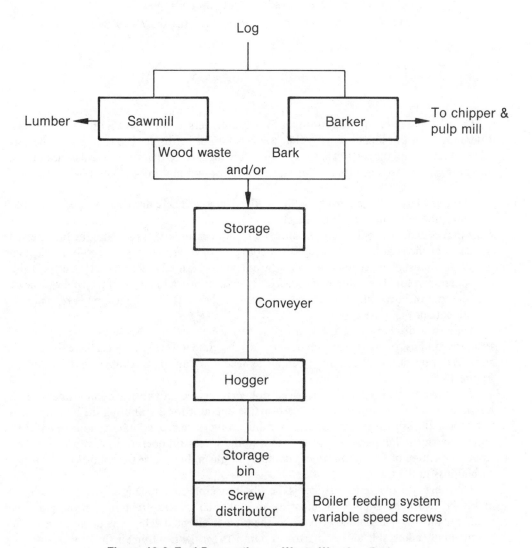

Figure 18-2 Fuel Preparation — Waste Wood or Bark

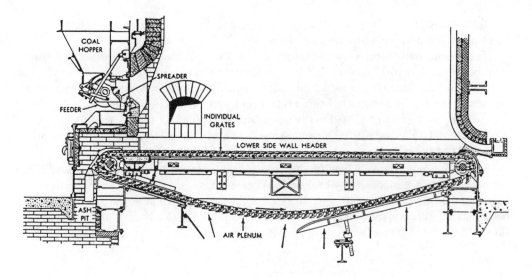

Figure 18-3 Travelling Grate Spreader Stoker

(From *Steam, Its Generation and Use*, ©Babcock and Wilcox Co.)

The multiple-retort underfeed stoker for larger boilers is shown in Figure 18-5. The basic difference from the single-retort is that the ash cannot be discharged at the sides because other retorts are on the sides. The grate is therefore inclined so that the ash will accumulate at the end of the retorts. In other respects, single-retort and multiple-retort underfeed stokers have similar characteristics.

The third basic type of stoker is the overfeed stoker. There are two subtypes: the chain grate and the vibrating grate. In the chain grate overfeed stoker, an endless chain type of grate moves under the bottom of the stoker coal hopper and drags coal into the furnace. The thickness of the coal bed is adjusted by a coal gate or dam that has a fixed but adjustable height. In almost all cases the coal gate adjustment is a manual function that is performed by the boiler operator. The volume of coal fed to the furnace is a function of the coal gate height and the rate of horizontal movement or speed of the grate. The coal flow is regulated by adjustments of the grate speed.

The ash is discharged at the rear of the furnace as the grate sections turn downward. Primary combustion air is fed from underneath the grate. Overfire air jets above the burning zone feed a portion of the combustion air as secondary air. This arrangement is shown in Figure 18-6.

The undergrate air section of overfeed stokers is divided into several compartments. Each has an adjustable air damper that is used by the operator for distributing the undergrate air to establish the fire line and the desired burning profile. The proper manual adjustment to the air compartment dampers is very important to the successful operation of overfeed stokers. Improper setting of the compartment dampers results in high excess combustion air or high carbon loss in the ash.

The other type of overfeed stoker is the vibrating grate stoker. This arrangement is shown in Figure 18-7. The grate is inclined in this stoker so that the material on the grate will gradually move toward the lower end, as the grate is vibrated. This type of stoker also uses the manually adjustable air compartment dampers for distributing air flow to the different grate sections.

Overfeed stokers disturb the burning mass of the coal to a relatively small extent. They are suitable for fuels that burn freely or are only weakly caking types. They are unsuitable for those coals that tend to fuse together while burning. Such action with an overfeed stoker would tend to shut off the flow of combustion air.

Of all the stokers, overfeed stokers have the highest storage of coal in the furnace prior the actual burning. The heat flow response to changes to fuel flow is therefore very slow. The major response is from a change in combustion air.

Mechanisms other than the stokers described are used for some applications involving industrial solid waste. For burning wood waste, a Dutch oven and conical pile combination is sometimes used. The new fuel drops to the top of the conical pile and air is blown around the pile. In the burning of bagasse, a highly sloped stationary grate with ash discharge at the bottom is used.

18-3 Special Stoker Control Problems

There are some special problems in the control of stoker firing that are unrelated to the control of combustion. The grate temperature must be kept low enough so that the grates will not be damaged. In some cases the stoker may be capabable of clean burning to a relatively low excess air value. Lower excess air increases the furnace temperature, which may cause grate damage. The grates must also be kept covered.

A hole in the fire on the grate causes combustion air to rush through the hole with insufficient air to other sections of the grate. When burning fuels that may be fed intermittently, the grate may become bare and exposed to the full radiant heat of the furnace. In such cases a large amount of combustion air must be fed through the bare grate to keep it cool. Another problem is the necessary manual adjustment of dampers for the proper distribution of the combustion air. The compromise between the characteristics of the load demand and the ability for stable response of the stoker to a heat flow demand is another consideration in the control of stoker-fired boilers.

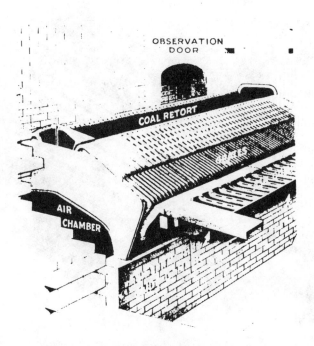

Figure 18-4 Single-Retort Underfeed Stoker

(From Detroit Stoker Co.)

Other coal burning problems that affect operator action and thus the ability of the control system to perform the entire control operation are listed below:

(1) Freezing of the coal in the bunkers or in storage.
(2) Bunker fires.
(3) Even distribution of coal on the grate.
(4) Grate air distribution.
(5) Overfeed stoker coal gate heighth.
(6) Coal clinkering on the grate.
(7) Incorrect type of coal for the stoker design.
(8) Excessive smoke.
(9) Excessive slagging of the ash.
(10) Incorrect or large variation in coal sizing and percentage of fines.

When burning waste fuels, other noncontrol problems may impact on the ability of the control system to perform its complete function.

(1) Fuel hang ups in storage bins and screw feeders.
(2) Size variation.
(3) Intermittent admission of fuel to the furnace.
(4) Large variations in fuel moisture content.
(5) Large variations in fuel unit Btu values.

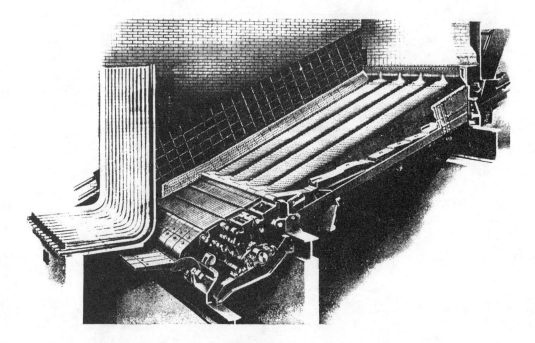

Figure 18-5 Multiple-Retort Underfeed Stoker

(From *Steam, Its Generation and Use*, ©Babcock and Wilcox Co.)

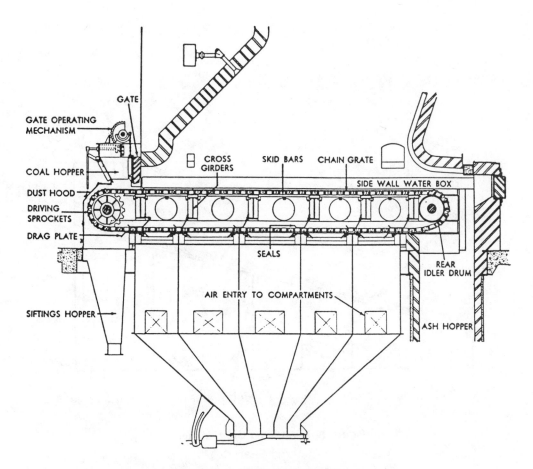

Figure 18-6 Chain Grate Stoker

(From *Steam, Its Generation and Use,* ©Babcock and Wilcox Co.)

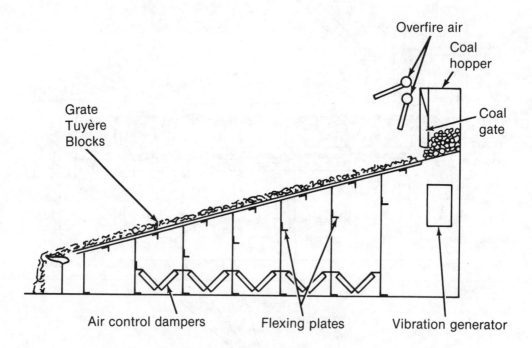

Figure 18-7 Vibrating Grate Stoker
(From Detroit Stoker Co.)

Section 19
Combustion Control for Stoker-Fired Boilers

Since a continuous measurement of the fuel burned for use in stoker-fired boiler control systems cannot be obtained, inferential measurement of the fuel input is necessary for metering types of systems. Parallel positioning systems can be used but with the same inherent inaccuracies and weaknesses as when they are used with liquid and gaseous fueled boilers.

19-1 Parallel Positioning Control Systems for Stoker-Fired Boilers

Parallel positioning systems can be applied to spreader stoker-fired boilers in the manner shown in Figure 19-1. The fuel control is usually a lever that rotates a shaft linked to the feeding mechanism of each of the spreader units. The number of the spreader units is determined by the capacity of the boiler, with each spreader unit feeding coal to a longitudinal strip of the grate. The position of the link is an approximate measure of the volume of coal. If the density of the bulk coal were constant, it would also be an approximate measure of the weight of fuel. Since a large percentage of the coal in a spreader stoker-fired boiler burns in suspension, the response characteristic of the fuel and air should be approximately parallel. This is shown graphically in Figure 19-1.

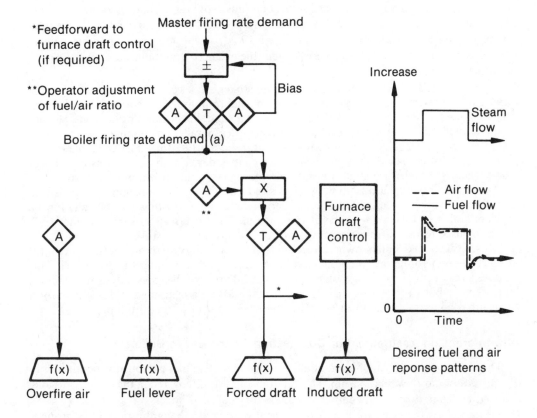

Figure 19-1 Parallel Positioning Control System for Spreader Stoker-Fired Boiler

The flow characteristic of the control signal vs. coal volume must be linearized and carefully matched to the flow characteristic of the required combustion air. This can be accomplished by using linkage angularity and cam shapes in the devices that position the combustion air control damper and the fuel control lever. The data for making these calibrations is derived from boiler combustion tests at three or more boiler loads spread over the operating range of the boiler.

Since all stoker-fired boilers are balanced draft boilers, the furnace draft must also be controlled, as discussed in Section 12. Overfire air requirements are somewhat unpredictable, although trends can be established over a period of time. The control for overfire air flow is normally a manual control when simple controls such as parallel positioning are used.

The rapid suspension burning of a large percentage of the coal makes spreader stoker-fired boilers more responsive to fuel input than combustion air input. For this reason the direct control signal should be applied to the fuel device. Any correction through use of manual control or a more sophisticated type of control system is then applied to the air flow control device.

If an underfeed stoker is used, the combustion process occurs over a period of time. The coal enters the furnace and is heated; distillation and separation of the components then must occur before combustion. There is thus a fuel storage that involves a mass of coal in various stages of burning in the furnace at all times. The primary heat response is from changes in combustion air flow.

As shown in Figure 19-2, the direct control signal is then applied to the air flow control device, with the secondary signal directed toward the stoker speed device. Any correction is then applied to the less responsive fuel flow. It is also desirable to add the time constant (j) in Figure 19-2 to smooth the stoker speed signal and thus avoid rapid changes in the coal flow, which have very little effect on the overall unit response. The relative changes of fuel and combustion air are shown graphically in Figure 19-2. As with the spreader stoker control system, furnace draft control is also necessary for the balanced draft boiler. The overfire air is shown as a manual control.

When overfeed stokers such as a chain grate stoker are used, the storage of fuel in the furnace prior to combustion makes the short term heat response of such boilers very insensitive to changes in fuel flow. Simple systems such as the parallel positioning are generally not adequate, and improved techniques are necessary. If a parallel positioning system can be used, the arrangement should be the same as that in Figure 19-2 for underfeed stokers. The primary signal should be directed to the air flow control device with a time constant (longer than that for the underfeed stoker) inserted in the signal to the stoker speed control.

The approximate relationship between desired flow changes of fuel and air flow is also shown graphically in Figure 19-2. These general relationships, as shown graphically for the spreader stoker in Figure 19-1 and for the underfeed and overfeed stokers in Figure 19-2, should be used for any arrangements of control systems for these stokers.

Most stoker-fired boiler owners recognize the desirability of more complete control systems in order to operate the boilers at lower excess air values and thus save fuel. This requires some form of metering system and/or the addition of flue gas analysis trim control loops. The inability to measure fuel Btu input requires that a metering control system for a stoker-fired boiler use some form of inferential measurement of fuel Btu input.

19-2 Inferential Measurement of Combustion Conditions in Boilers

The most common form of inferential relationship that defines combustion conditions is the steam flow/air flow relationship. This relationship was discovered in the early 1900s and patented as the first practical combustion guide for coal-fired boilers.

Steam flow is the usual measurement of heat output. If the combustion air is proportioned to the steam flow (heat output), certain combustion conditions will exist, thus producing a certain boiler efficiency. The steam flow heat output can be divided by the efficiency to produce heat input. Steam flow is thus an inferred measurement of heat input and when correlated to combustion air flow implys a relationship between fuel input and air flow. There are known specific limitations of the use of this technique:

(1) Steam flow is not proportional to heat input when load is changing and over or underfiring is necessary to satisfy the heat storage requirements of the boiler.

(2) Changing the excess air changes the boiler efficiency and thus changes the inferred relationship between steam flow and heat input.

(3) The cleanliness of the boiler (soot on the fireside surfaces) changes the boiler efficiency and the relationship between steam flow and fuel heat input.

(4) Changes in feedwater temperature or boiler blowdown cause the relationship between boiler steam flow and heat output to change. The result is a change in the steam flow/heat input relationship.

(5) Operating at other than design steam pressure and steam temperature produces errors in the steam flow measurement. The result is a change to the measured steam flow/heat input relationship.

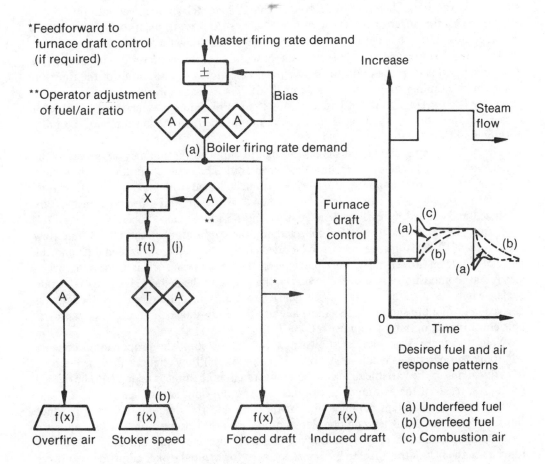

Figure 19-2 Parallel Positioning Control System for Underfeed or Overfeed Stoker-Fired Boilers

While the steam flow/air flow relationship is not a precise measurement of combustion conditions, the errors above are not large errors. The steam flow/air flow relationship has been used very successfully as a combustion guide and applied as an automatic control input for approximately 70 years.

The steam flow/air flow relationship is a field calibration based on the results of a series of combustion tests over the operating range of the boiler. The air flow mechanism is adjustable so that its calibrated output matches the output of the steam flow measurement when the desired combustion conditions are obtained. The air flow measurement is therefore a relative one and does not have an output calibrated in terms of pounds or cubic feet. In this respect, the calibration method is the same as the relative air flow calibration used when burning liquid or gaseous fuel.

19-3 Parallel Positioning Control Systems with Steam Flow/Air Flow Readjustment

Because of the steam flow/air flow limitations listed above, the steam flow/air flow relationship is recommended for use as a corrective or trim control. During a rapid load increase, close adherence to the desired 1:1 relationship of steam flow to air flow will cause a reduction in excess air. The result may be smoke, higher furnace temperature, and potential damage to the grates. For this reason, there should be a deviation from the 1:1 relationship during load changes in order to maintain a more constant excess air level. The steam flow/air flow influence should therefore be delayed so that it is not felt or averaged into the control system during the under or overfiring of load changes. An arrangement for the control of a spreader stoker-fired boiler that meets these conditions is shown in Figure 19-3.

The boiler firing rate demand (a) is the direct signal to the stoker lever driving device (b). This signal also acts as the input to function generator (c) with the output of the function generator positioning the overfire air damper (d). The function installed in this function generator is an average function based on observation of required values over a period of time. The operator has available a bias of the output of this function for more accurate positioning of the overfire air.

The boiler firing rate demand signal, through its input to the summer (e), serves as the initial and primary signal for controlling the forced draft. The initial calibration of the system is that of a parallel positioning system. Based on data collected from boiler combustion tests, the control signal vs. fuel flow and control signal vs. desired air flow relationships are linearized and aligned, using linkage angularity and positioner cam shapes.

During these tests, data is also collected for the calibration of the relative air flow measurement (f). This measurement is calibrated so that the value of the signal (f) and the value of the steam flow signal (g) are equal over the entire load range. This ensures that when steady state combustion conditions are correct, the output of the delta (h) is equal to 0. Upon a change in boiler load that would cause signal (h) to change in a plus or minus direction, it is delayed in its input to summer (e) to allow a return to steady-state conditions before appreciable effect on the output of summer (e).

Note that there is no integral shown in this control loop. The proportional offset is adjustable through the gain adjustment on the steam flow/air flow relationship. The tuning is not critical due to the variable time constant adjustment in the time function (i). The overall tuning is a result of the gain in delta (h) plus the time constant adjustment (i).

Within limits, different combinations of the gain and the time constant will produce satisfactory results. A higher gain in order to produce a lower proportional offset requires a longer time constant. If a proportional offset exists during steady-state operation, the operator can use the fact of the offset to recognize the need for manual stoker adjustment or other action resulting from changing coal quality or boiler fireside fouling, etc.

Assuming that the system logic can accept signals with both positive and negative values, the output of the delta (h) and the time function (i) are 0 or nearly so under steady-state operation. The signal (a) should have a gain of 1.0 at its input to summer (e). With a 0 signal from time function (i), the bias adjustment of summer (e) would be set at 0. If the system can operate only on positive value signals then a plus 50% bias would be added at the output of delta (h) and a minus 50% bias adjusted in the summer (e).

A parallel plus steam flow/air flow readjustment control system for underfeed and overfeed stoker-fired boilers is shown in Figure 19-4. There are two essential differences between this control arrangement and that for a spreader stoker. The signal from summer (e) is shown connected to fuel with the direct signal (a) connected to the forced draft. This is a result of the different basic response characteristics to fuel and air flow changes. The other change is the addition of the time function (j). This allows the fuel response to be adjusted to a desired fuel response curve as shown in Figure 19-2. All other functions of the control system and their calibration are the same as that for the spreader stoker system of Figure 19-3.

In addition to the operation of the control sytem, operator action is required. The operator must set the fuel bed thickness with the coal gate height. The operator must set air flow distribution with the compartment dampers. In the control arrangement shown, over-fire air is controlled as a function of boiler firing rate, but the operator must bias the setting if

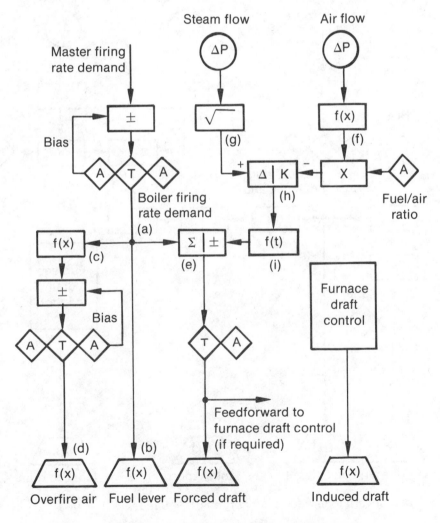

**Figure 19-3 Parallel Positioning Control System with
Steam Flow/Air Flow Readjustment for Spreader Stoker-Fired Boiler**

necessary. A furnace draft control loop is also necessary with the balanced draft boiler. When blowing soot, the operator may need to increase the furnace draft set point to a more negative value. He may also need to increase the combustion air to fuel ratio during the soot blowing period. Therefore, the total control system is the result of actions based on visual observation by a skilled operator and an imperfect automatic control arrangement.

19-4 Series Ratio Control Systems for Stoker-Fired Boilers

Another form of control system that is applied to stoker-fired boilers is known as the series ratio system. This type of system makes use of the steam flow/air flow relationship without the direct linkage of a parallel positioning system. The linkage of the fuel and air control is provided through the process. The net result is control action that is somewhat similar in its overall pattern to that of the parallel positioning plus steam flow/air flow readjustment. Figure 19-5 shows such a system for a spreader stoker-fired boiler.

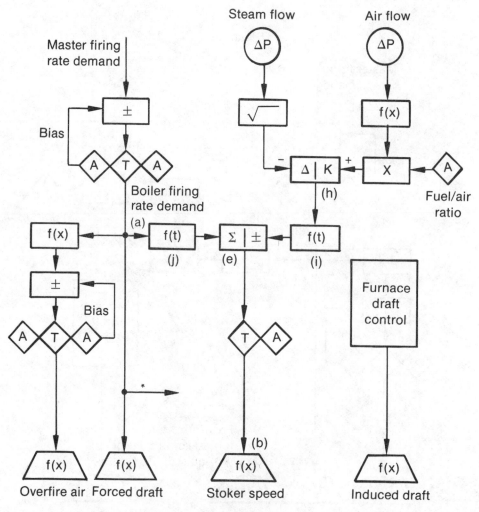

*Feedforward to furnace draft control (if required)

Figure 19-4 Parallel Positioning Control System with Steam Flow/Air Flow Readjustment for Underfeed and Overfeed Stoker-Fired Boilers

The boiler firing rate demand (a) directs the fuel feed through adjustment of the fuel lever. In the arrangement shown, overfire air is controlled in the same manner as in the parallel-plus-readjustment type of system. The boiler firing rate demand is a measure of the requirement for steady-state steam flow plus or minus the need for adjustment to boiler energy storage. This signal need not be connected to the combustion air flow control if there can be other signals available that are analogous to the firing rate demand signal.

In the system shown in Figure 19-5, the firing rate demand signal (a) controls the fuel lever directly as in the parallel and parallel-plus-readjustment type of system. The signal from the proportional-plus-derivative function (b) is analogous to the firing rate demand signal (a). The steam flow portion is measured steam flow with the proportional-plus-derivative function representing the rate at which the boiler energy storage should be adjusted.

In tuning the combined function (b), the proportional gain (h) is adjusted to provide the magnitude of the over-or underfiring necessary, with the derivative time (k) representing the time period over which the over- or underfiring takes place. The result is a proper set point for the combustion air flow proportional-plus-integral controller (c). The direct input from steam flow into the combining summer (b) should be set at a gain of 1.0 in order that steam flow and air flow will be together at all loads. Since the derivative action takes place only during a load change, the steam flow/air flow relationship is in 100% control of the fuel during the steady state condition of boiler load.

Because the firing rate demand signal (a) is not used in the control of combustion air, the fuel (g) and forced draft (f) devices do not require the careful alignment calibration of the parallel-plus-readjustment type of system. They should be linearized so that control response will be equal at all loads. Since full reliance is on steam flow/air flow during steady-state

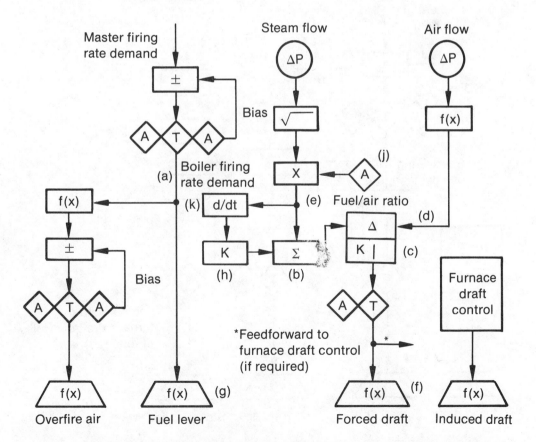

Figure 19-5 Series Ratio Control System for Spreader Stoker-Fired Boilers

operation, the relative air flow measurement (d) must be carefully calibrated to match the steam flow measurement (e). As described earlier, this calibration is based on a fuel/air ratio setting (j) of 1:1 and combustion tests at three or more boiler loads over the boiler load range.

A series ratio control system for an underfeed or overfeed stoker-fired boiler is shown in Figure 19-6. In this arrangement, since the boiler is much more responsive to air flow than fuel flow, the firing rate demand signal (a) is the set point of the air flow controller (h) with feedback from the relative air flow measurement (d). Assuming steady-state operation and a fuel/air ratio setting (j) of 1:1, the steam flow measurement (e) directly positions the stoker speed (g). During a load change, the time function (i) smooths and delays the change in fuel flow to the stoker as shown on the desired fuel response curve in Figure 19-2.

The characteristic of this system is that variations in coal quality or other operational factors affecting the system operation will create a steady-state offset in the steam flow/air flow relationship. As in the case of using proportional-only readjustment in the parallel-plus-readjustment system, the operator can use the offset indication to provide additional intelligence to his part of the overall control system operation. This type of system will produce an offset based on a steam flow/air flow gain of 1.0 instead of the adjustable gain of the other type of system.

The system is aligned and calibrated by matching the stoker speed control (g) characteristic to the air flow measurement signal (d). Since the stoker speed vs. coal flow characteristic is essentially linear, the calibrated air flow signal will also be nearly linear. As before, this

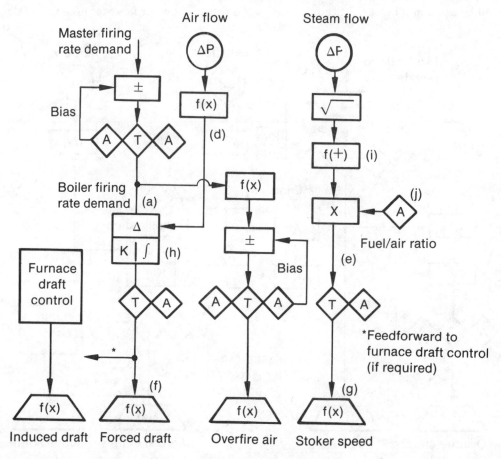

Figure 19-6 Series Ratio Control System for Underfeed and Overfeed Stoker-Fired Boilers

calibration is based on boiler combustion tests at three or more loads and includes the variable excess air vs boiler load relationship.

19-5 Applying Flue Gas Analysis Trim Control to Stoker-Fired Boilers

Flue gas analysis trim control for stoker-fired boilers is applied by substituting a block of trim control functions for the manual fuel/air ratio control shown in Figures 19-1 through 19-6. It should not be applied indiscriminately where factors other than flue gas analysis limit the reduction of excess air. These limiting factors to the reduction in excess air are discussed in Section 16-6. Factors that particulary apply to stoker-fired boilers are:

(1) Carbon in the refuse
(2) Slagging, difficulty in ash handling
(3) Smoke and particulate carryover
(4) Burning of grates from high furnace temperatures, other furnace and stoker maintenance

If smoke is the limiting factor, the trim control in Figure 16-9 can be used. If any of the above except smoke is the limiting factor, then the % oxygen trim control of Figure 16-6 may be used. A restriction is that the effect of the limiting factor should be included in the development of the curve for load vs. desired % oxygen. A trim control arrangement of % oxygen and ppm CO in accordance with Figure 16-6 would be permissable if the lower edge of the % oxygen band would not cause the breaching of one of the limiting factors. If there are no other limiting factors and the only limit to excess air reduction is the optimization of boiler efficiency, % oxygen and ppm CO can be used without reservation as the final trim control index.

Note that the use of % carbon dioxide plus ppm CO has not been discussed as a trim control method. While % oxygen can be used alone, % carbon dioxide cannot. For any particular fuel a change in % oxygen is always accompanied by a change in % carbon dioxide. When excess air is reduced and carbon monoxide appears, however, the relationship is distorted and % carbon dioxide becomes an unreliable index. If, however, ppm CO is used along with % CO_2, the ppm CO furnishes the intelligence for determining whether or not the % CO_2 is a reliable index. Applied in such a manner, % CO_2 plus ppm CO can be used interchangeably with % O_2 plus ppm CO. A change in the hydrogen/carbon ratio of the fuel changes the excess air/% CO_2 relationship while the same change has a very minor effect on the excess air/% O_2 relationship. In addition % O_2 analysis is more precise and less costly than a % CO_2 analysis, therefore, the natural choice is % O_2.

The application of any trim control arrangement should be carefully examined to determine its potential for breaching any of the limiting factors to reducing excess air. For example, one particular combination of flue gas analyses for use in stoker trim control loops is % CO_2 plus ppm CO. The particular application describes the application of ppm CO to the control of overfire air with the undergrate air used to optimize the % CO_2. It can be shown analytically that unless some other limiting factor is used to stop the % CO_2 optimizing action, the result will be an excessive amount of carbon in the refuse.

When applying the flue gas analysis trim control to a particular control arrangement, the tuning of the trim control will be affected by the control elements downstream from the point at which the trim control enters the loop. Figure 19-7 shows the application of trim control to the parallel-plus-SF/AF system of Figure 19-3. Assume that the trim control is a % oxygen system as shown in Figure 16-6. The integral time setting of the trim control must always be significantly slower than the time function setting of item (i). The overall system will show instability due to interaction between these functions as these time settings are moved closer together.

The interaction is similar to that of a cascade loop, where the tuning time constant of the downstream controller integral action should be an approximate order of magnitude faster than that of the upstream controller. The practice should be to adjust time function (i) for proper action with steam flow/air flow control. The % oxygen controller integral setting should then be adjusted for overall system stability. The limitation above also applies to the addition of flue gas analysis trim control to underfeed and overfeed stoker-fired boilers. This arrangement is shown in Figure 19-8.

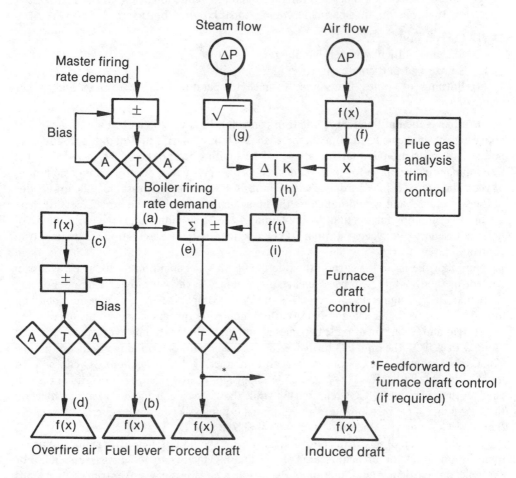

Figure 19-7 Parallel Positioning Control System with Steam Flow/Air Flow Readjustment and Flue Gas Analysis Trim Control for Spreader Stoker-Fired Boilers

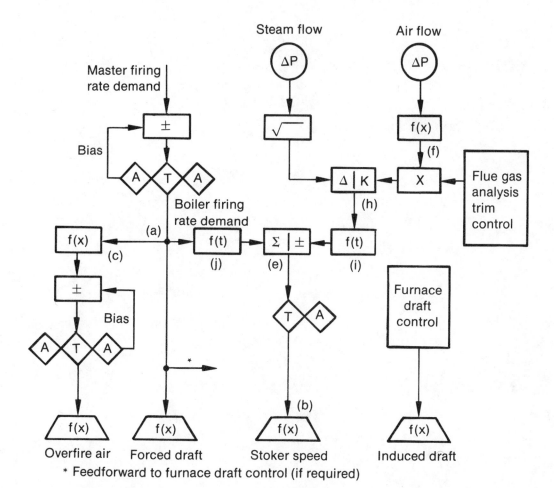

Figure 19-8 Parallel Positioning Control System with Steam Flow/Air Flow Readjustment and Fuel Gas Analysis Trim Control for Underfeed and Overfeed Stoker-Fired Boilers

Section 20
Pulverized Coal Burning Systems

In pulverized coal-fired boilers the coal is ground by a coal pulverizer to a fine powder and blown into the furnace. From a control standpoint the burning of pulverized coal is similar to that of a gaseous fuel with rapid heat response to changes in fuel flow and minor response to changes in air flow. A typical arrangement of the unit pulverized coal system is shown in Figure 20-1. A pulverized coal burner is similar to a gas or oil burner with coal transported on a stream of primary air and with secondary air, the main body of combustion air, added at the burner. In the unit system shown, which is the present day system of general use, the coal is ground as it is used. Depending on the type of pulverizer used, there is more or less pulverized coal storage in the pulverizer prior to burning. Each pulverizer serves more than one burner. This requires that orifices are placed in the coal-primary air pipes to each burner so that the coal flows to the different burners are balanced.

A typical unit pulverizer system contains four basic elements:

(1) The coal feeder
(2) The pulverizer and classifier
(3) The primary air flow supply or exhauster fan
(4) The pulverized coal drying system

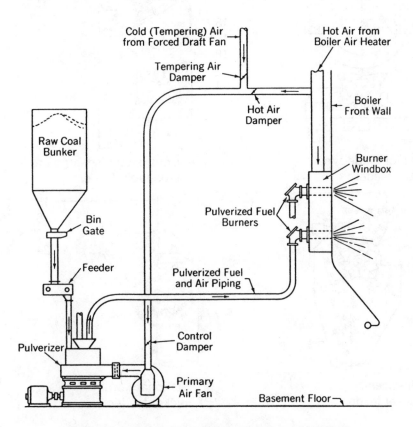

Figure 20-1 Direct-Firing System for Pulverized Coal

(From *Steam, Its Generation and Use*, ©Babcock and Wilcox Co.)

20-1 The Coal Feeder

The coalfeeder feeds the raw coal from the overhead bunkers to the pulverizer. The coal should be fed to the boiler in approximate synchronization with the rate at which it is burned in the furnace. There are two basic types of feeders: the volumetric feeder and the gravimetric feeder.

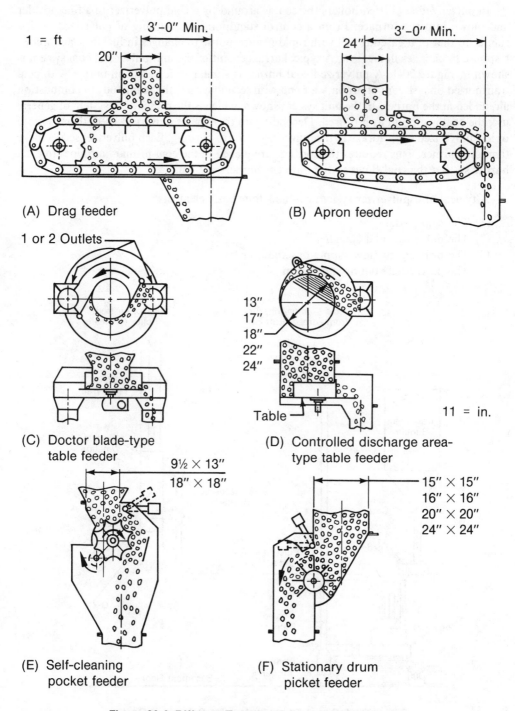

(A) Drag feeder

(B) Apron feeder

(C) Doctor blade-type table feeder

(D) Controlled discharge area-type table feeder

(E) Self-cleaning pocket feeder

(F) Stationary drum picket feeder

Figure 20-2 Different Types of Volumetric Coal Feeders

(From *Stock Equipment Co.*, technical paper)

A volumetric feeder feeds coal by volume. As the speed of the feeder increases, the volumetric rate of coal feed increases. Several different types of volumetric feeders are shown in Figure 20-2. The weakness of the use of the volumetric feeder is that the bulk density of the coal varies, resulting in variation in the weight of coal fed and thus variation in total Btu to the furnace. This is an appreciable variation since the the bulk density for most coals (different for lignite) ranges from approximately 38 lbs/ft^3 to 49 lbs/ft^3 for a moisture reduction range of 6 percent. This represents a plus or minus 18.3 percent variation in total Btu feed for the same volume of coal.

The gravimetric feeder is both a weight flow meter and a feeder that feeds coal to the pulverizer on the basis of coal weight. The weight of coal feed to the boiler is directly proportional to the control signal to the coal feeder. A typical gravimetric feeder arrangement is shown in Figure 20-3. A belt weighing device inputs to a servomechanism that adjusts the leveling bar to maintain 100 lbs. of coal on the belt at all times. The weight of the coal fed is thus directly proportional to the speed of the belt. The speed of the belt is thus a coal weight flow rate.

A newer type of gravimetric feeder uses a fixed-height leveling bar and electronic load cell for weighing the belt. The coal feed is the product of belt weight and belt speed. With the moisture vs. bulk density relationship above and for a constant coal weight feed, the total Btu input variation is plus or minus 3%, the variation in the coal moisture content. A typical

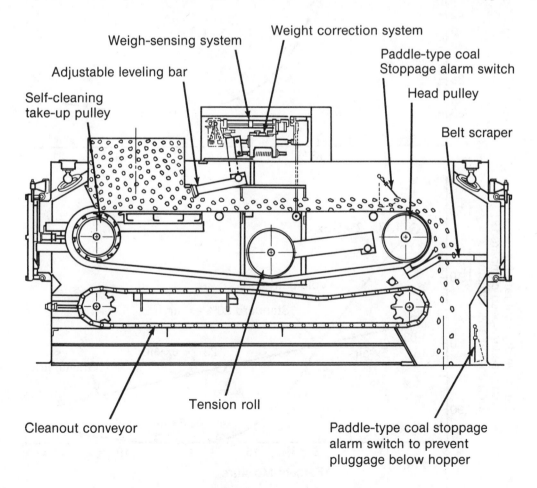

Figure 20-3 Gravimetric Coal Feeder, Belt Type

(From *Stock Equipment Co.*, technical paper)

comparison between a gravimetric and volumetric feeder is shown in Figure 20-4. If the combustion air does not change, these variations in Btu feed to the furnace create changes in total air for combustion and affect both steam pressure and temperature.

The basic benefit of a gravimetric feeder is that it compensates for variations in bulk density that arise from changes in coal moisture. The gravimetric feeder does not compensate for Btu input variation that results from changes in ash content of the coal. While an increase in moisture content causes the bulk density to decrease, an increase in ash content causes it to increase. If the coal is of poor quality with wide variation in ash content, the effect of such variation on total Btu input to the furnace can be greater with a gravimetric feeder than with a volumetric feeder.

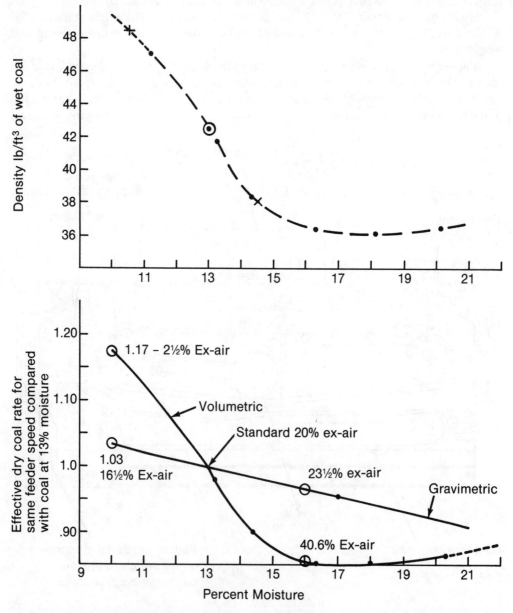

Figure 20-4 Comparison of Excess Air Effect for Gravimetric and Volumetric Coal Feeders

(From *Stock Equipment Co.*, technical paper)

20-2 The Pulverizer and Classifier

The pulverizer is an electric motor-driven rotating mechanical grinding mechanism that crushes the coal between rotating balls and a race as in Figure 20-5, between a roller and a bowl as in Figure 20-6, between tumbling steel balls in a cylinder partially filled with the balls as in Figure 20-7, and various types of hammer or impact grinding mills as in Figure 20-8. Some coals require a particular type of grinding mill. A good example is that ball mills are more successful in the pulverization of very hard or abrasive coals such as anthracite or meta-anthracite coals.

In some pulverizers, the fineness of the grinding is determined by spring force on the grinding surface devices. In other pulverizers, centrifugal action applies greater contact force between the grinding surfaces as rotational speed is increased. The classifier rejects the larger particles to the pulverizer. The ground coal size is also determined by the size that is rejected by the classifier. In a typical installation the coal fineness is in the range of 70 to 80% through a 200-mesh screen.

Coal is continuously fed to the pulverizer and primary air is continually passed through the pulverizer to pick up the coal that has been pulverized. The amount of pulverized coal lifted from the pulverizer depends on the fineness of the pulverized coal, the level of coal in the pulverizer, and the square of the primary air flow. The turndown control range for a pulverizer is approximately 2:1. Typically, excessive coal moisture or hardness may reduce a pulverizer's capacity to as low as 50% of its rated capacity.

Because of these limitations on turndown and capacity, plus the fact that pulverizers are built in standard sizes, a pulverized coal boiler installation will usually require multiple pulverizers. Because the pulverizer is a heavy duty piece of grinding machinery that is subject to considerable wear and maintenance requirements, a typical installation will include an additional installed spare pulverizer.

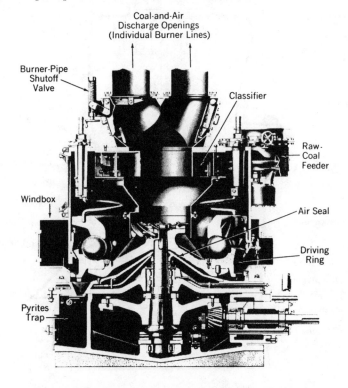

Figure 20-5 Medium-Speed Pulverizer, Ball-Race Type

(From *Steam, Its Generation and Use*, ©Babcock and Wilcox Co.)

As the coal is ground and lifted in the air stream, it first passes through the classifier. In the classifier, centrifugal action is imparted to the stream through static or rotating elements. The larger and heavier particles are thrown to the outside and returned to the pulverizer for further grinding.

From a control standpoint, there are different ways in which the same pulverizer can be controlled. There are, however, some generally accepted methods by which particular pulverizers are controlled. The differences are based on the manufacturer's control philosophy, whether the pulverizer is pressurized or under negative pressure, the type of pulverizer (which determines the amount of pulverized coal storage in the pulverizer), and whether the pulverizer is operated at a fixed or variable speed.

20-3 The Primary Air Flow or Exhauster Fan and the Coal Drying System

The source of the air flow that carries the pulverized coal stream from the pulverizer to the burners is a primary air fan or an exhauster fan. If the pulverizer is under air pressure, the fan pumps clean air to the pulverizer and then to the burner. If the pulverizer is a type that is operated under negative pressure, the exhauster fan is between the pulverizer and burners.

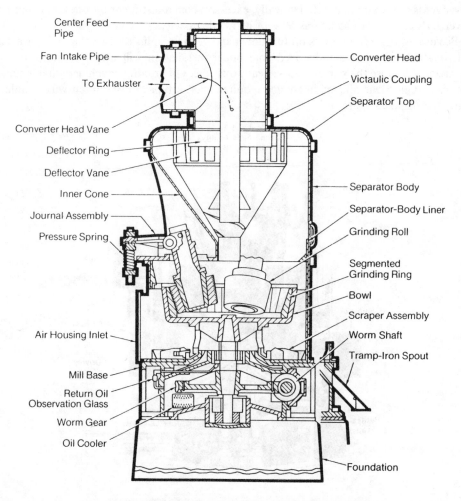

Figure 20-6 Medium-Speed Pulverizer, Roller-Bowl Type

(From *Fossil Power Systems*, ©Combustion Engineering, Inc.)

Exhauster fan systems can also be used with pressurized pulverizers. An exhauster fan pumps the primary air-pulverized coal mixture from the pulverizer to the burners. A separate exhauster fan is necessary for each pulverizer. Because of the abrasive characteristic of the pulverized coal, an exhauster fan must be of rugged construction and of simple and less efficient design to stand up to the wear on the fan parts.

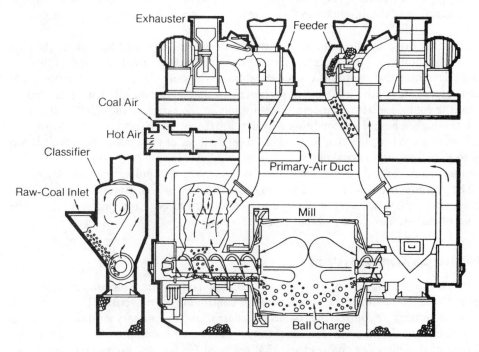

Figure 20-7 Slow-Speed Conical-End Pulverizer with Segregated Ball Size Indicated

(From *Fossil Power Systems*, ©Combustion Engineering, Inc.)

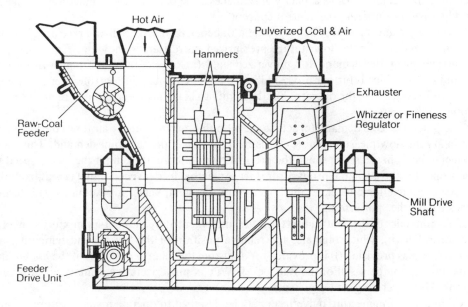

Figure 20-8 High-Speed Pulverizer, Impact Type

(From *Fossil Power Systems*, ©Combustion Engineering, Inc.)

For pressurized pulverizers the primary air flow for both types of fans is controlled on the clean air side of the pulverizer. The air supplied to the pulverizer is a mixture of hot, preheated air from the boiler air preheater and relatively cold ambient temperature air called tempering air. While the amount of coal lifted from a pulverizer at constant coal level is approximately proportional to the square of primary air flow, changing the coal level also has a very significant effect on the coal lifted in the air stream. The change in pulverizer coal level is a function of the difference in flow rate between raw coal input and pulverized coal output.

The coal is dried by the primary air stream mixture of cold and hot air. If the coal contains more moisture, then more hot air and less tempering air are used in the mixture. A temperature sensor that measures the temperature of the pulverized coal-air mixture feeds back to control the relative amounts of preheated air and tempering air. The controlled set point of the coal-air mixture temperature is usually in the range of 140–160° F. If the set point is too high, a hazardous condition may develop from pulverized coal that is too dry. If the set point is too low, the coal may build up on surfaces of the system, because it is too damp to flow freely in the coal pipes.

Just as in handling all volatile fuels, there is some hazard in pulverizing and drying the coal. The potential problem is that of fires in the pulverizers or coal pipes. An emerging fire detection technique is to analyze the mixture of pulverized coal and primary air for the presence of carbon monoxide or other combustible gases.

20-4 Pulverizer Control Systems

A block of control logic is needed for each individual pulverizer. This control logic block contains all the modulating control functions for controlling the pulverizer. These functions are the coal-air mixture control, the coal feeder control, and the primary air flow control. As indicated above, different control schemes are normally used for different types of pulverizers. Some of the blocks of control logic that are used for the well known pulverizers are shown and discussed below. Digital logic that is necessary for startup and shutdown sequencing and for safety interlocks is not shown. Since the pulverizer manufacturers normally guarantee such capabilities as capacity and fineness, the particular manufacturer should be consulted before finalizing the control concept.

A typical arrangement for the control of a Babcock & Wilcox ball and race or roller and race pulverizer and similar low storage pressurized pulverizers is shown in Figure 20-9. In this arrangement the demand for pulverized coal (a) originates as the boiler firing rate demand signal, which is further processed in the boiler combustion control system. Through the bias adjustment (b), the operator has the ability to bias the pulverizer load relative to that of other pulverizers.

The pulverizer coal demand (c) acts as the set point for the primary air flow controller (d). A primary air flow measurement (e) feeds back to satisfy the set point demand. This measurement also is processed through the proportional function (f) to limit the feeder speed to that proper for the available primary air flow. An actual feeder speed signal is sent from the pulverizer control logic block to the boiler control system, where it is summed with feeder speeds from other pulverizers.

The remainder of the control functions are implemented by the coal-air mixture temperature controller (h), which regulates the relative position of the hot air and tempering air dampers. A bias provision (i) is provided for adjusting the relative positions of the hot air and tempering air dampers. All of the Babcock & Wilcox pulverizers use primary air fans with the pulverizer operating under pressure.

Combustion Engineering pulverizers may be designed to operate under pressure or may be designed as draft systems. In Figure 20-10 the control logic for a pressurized system is shown. The basic pulverizer demand signal (a) can be biased by the operator using the bias function (b). This provides a means of balancing pulverizer loads and smoothly bringing the pulverizer on-line or taking it off-line. The resulting signal (c) positions the coal feeder directly and is further processed in proportional-plus-bias function (d) to a proper feedforward signal for the primary air flow. In this arrangement the primary air flow controller (f) is at a fixed set point as adjusted by the operator and is satisfied by the primary air flow feedback (g). The output of the controller is directed to the hot air damper and as a feedforward signal to summer (k) and then to the cold air damper.

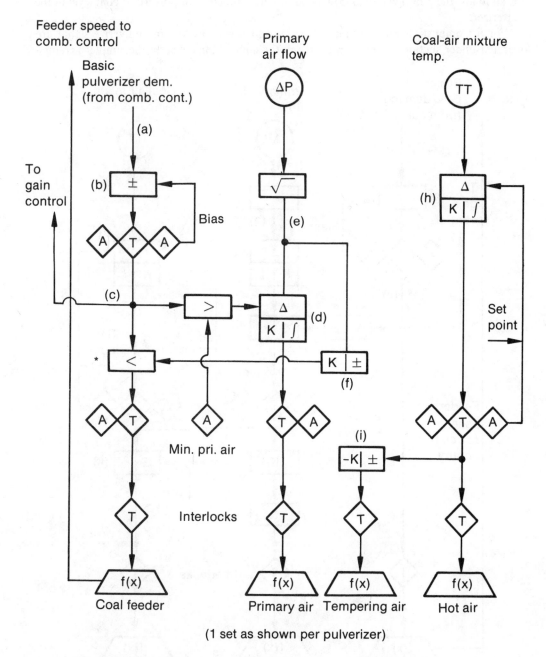

Figure 20-9 Control System for Babcock and Wilcox Low Storage Pulverizers

The coal-air mixture temperature controller (i) regulates the temperature by changing the distribution of hot and cold air. As the temperature drops and more drying is needed, the hot air damper opens further and cold air damper closes a like amount. In essence, both dampers control primary air flow and both dampers control coal-air mixture temperature. The tuning procedure for these control loops is that for typical temperature and flow control loops.

A signal of actual feeder speed is sent from the control logic block shown to the boiler control system. Note that since under steady-state conditions the primary air flow is constant for all loads, the pulverized coal flow is essentially a function of pulverized coal level in the pulverizer.

For a Combustion Engineering pulverizer and other low volume pulverizers that are operated under suction rather than pressure, the control logic can be the simpler arrange-

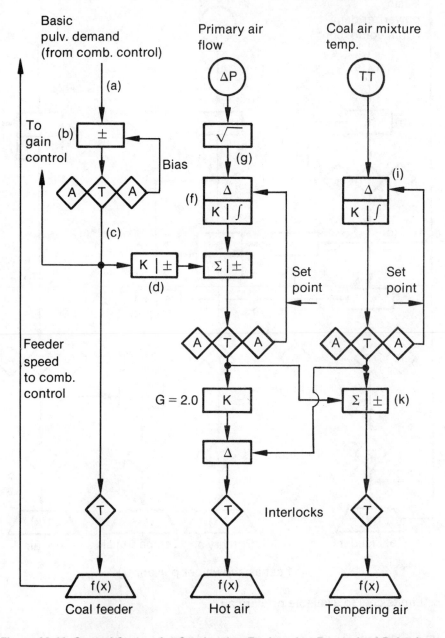

Figure 20-10 Control System for Combustion Engineering Pressurized Pulverizers

ment shown in Figure 20-11. In this case, the pulverizer demand signal (c) positions the feeder speed and the primary air (exhauster) damper in parallel. The tempering air is introduced through a balanced draft damper, which opens as a suction is created at the pulverizer air inlet. The coal-air mixture temperature controller (i) is a simple feedback controller that controls only the preheated air that mixes with the tempering air.

A typical Foster Wheeler pulverizer is of the ball mill type shown in Figure 20-7. Pulverizers of this type may have several minutes' storage of pulverized coal. At constant coal level they are relatively insensitive to immediate changes in the raw coal feed rate but have a significant amount of ground coal available for immediate response to load demand. Figure 20-12 shows how such a pulverizer may be controlled.

The coal level (a) in such pulverizers is sensed with an air bubbler or air leak off probe. As the level changes, the coal level controller (b) signals a change in the coal feeder speed to satisfy the change in demand for coal. The demand signal for pulverized coal (c) is the set point for the primary air flow controller (d), which positions the damper on the exhauster fan. The classifiers are separate devices between the exhauster fan and the pulverizer. The classifier pressure differentials (e) are summed and used as the pulverized coal flow feedback to the combustion control system.

As the change in the exhauster damper changes the draft in the pulverizer, the draft controller (f) changes the tempering air flow. Changes in the coal-air mixture temperature cause the temperature controller (g) to change the position of the hot air damper.

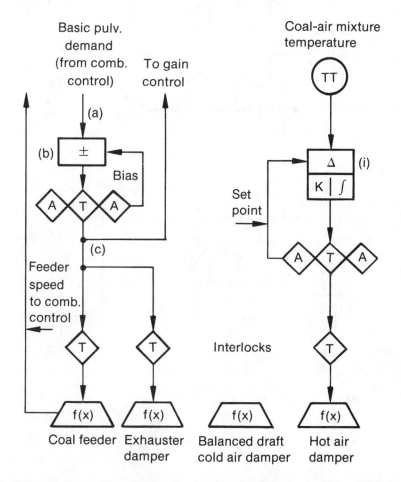

Figure 20-11 Control System for Combustion Engineering Negative Pulverizers

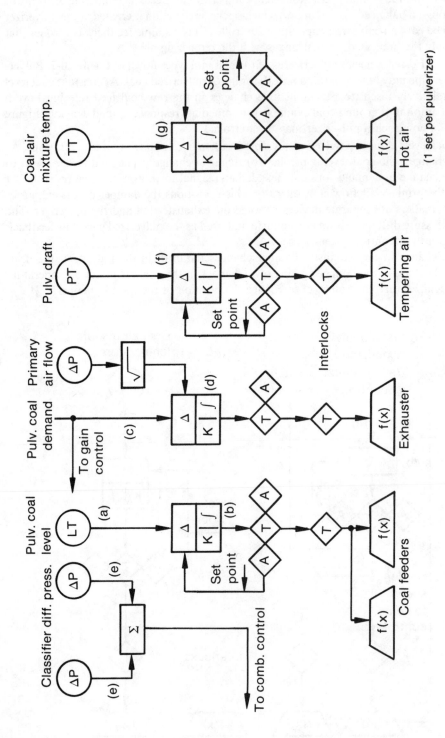

Figure 20-12 Control System for Foster Wheeler Ball Mill Pulverizers

Another variation for controlling this type of pulverizer eliminates the weakness of the classifier differential as a pulverized coal flow measurement. This arrangement is shown in Figure 20-13. The exhauster flow damper is positioned directly by the pulverizer demand signal (c). The coal level control of the pulverizer is a feedforward system. The signal to the exhauster is fed forward to the summer (h) as the preliminary control signal for the coal feeder, with the coal level controller (b) acting as a trim control. The input of the coal feeders is then in phase with the pulverized coal output from the exhausters.

If the feeder input causes a temporary rise in the coal level, the increase in the level causes the exhausters to deliver more pulverized coal. The feeder speed summation (j) is then gained by the level error (i) in multiplier (g) before being used in the boiler control system. This shows that the exhauster coal flow is different from the raw coal feed because the level is in error. The calibrated actual feeder speed summation (g) is then the feedback to the combustion control system. The feedforward variation of the coal-air mixture and draft control shown here can be used with either of the basic control arrangements.

The Riley ball mill pulverizer is very similar to that of Foster Wheeler, but the control philosophy appears to be quite different as shown in Figure 20-14. In the Riley system the coal feed can be controlled by pulverizer level, as in the Foster Wheeler system, by using the pneumatic probe type of pulverized coal level measurement. Riley also provides an alternative system (called Powersonic R) that uses inputs of sonic level, the sound level generated by the grinding process, and kW rate to control the raw coal feeders. The flow rate of pulverized coal is determined by controlling the pulverizer differential (e) according to the demand for pulverized coal. The summation (e) of the pulverizer differential pressures can then be used as a rough pulverized coal flow feedback to the boiler combustion control system. The damper controlling the differential is called the mill rating damper.

The classifier-to-furnace differential is controlled to a constant value by the action of controller (g). The output of this controller regulates the mill bypass dampers to bypass a portion of the primary air flow around the pulverizer. Since this differential is affected primarily by the primary air flow, the primary air flow is approximately a constant value for all coal loads with a variable content of pulverized coal in the mixture. The flow through the pulverizer is a part of the flow through the classifiers and thus contributes to the classifier-to-furnace differential pressure. The control signal to the mill bypass dampers is therefore used as an input to summer (f) in order to act as a feedforward signal to the mill rating damper. As in the other systems, the coal-air mixture temperature, which is not shown, controls the distribution of hot and preheated air in the primary air stream.

The control arrangements shown are for the larger and more complex pulverizers. If simpler or smaller pulverizers are used, simpler control arrangements may be adequate.

All of the pulverizer control arrangements shown, or which may be applied, require one set per pulverizer. The boiler systems usually require three or more pulverizers for full load operation (two or more operating and one spare), although they may be operated with a fewer number at low loads. The boiler combustion control system should compensate for the number of pulverizers in use by automatically changing the gain of the pulverizer demand signal before it reaches the individual pulverizer controls.

All of these pulverizers or mills have some inherent time lag between the raw coal feed and the output of pulverized coal. This has no effect during steady-state operation, but can create control problems due to delay of fuel response when boilers have rapid changes in load. The normal practice is to add derivative response in the combustion control system ahead of the pulverizer control functions in order to improve the response.

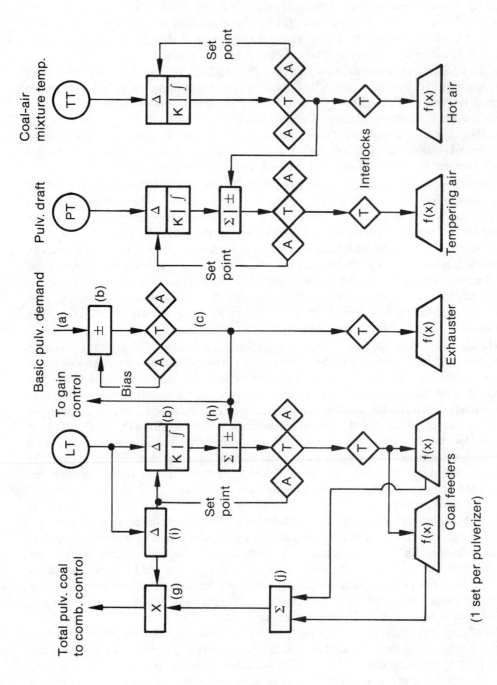

Figure 20-13 Control System for Ball Mill Pulverizers Using Gravimetric Coal Feeders

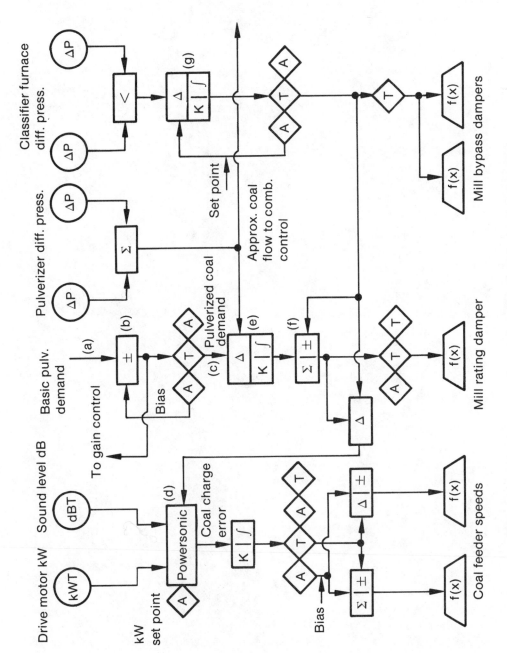

Figure 20-14 Control System for Ball Mill Pulverizer with Powersonic R

Section 21
Combustion Control for Pulverized
Coal-Fired Boilers

Combustion control systems for the firing of pulverized coal should be viewed as similar to those for gaseous or liquid fuel. The fuel flows in the current of primary air as a fluid, and the combustion and associated heat release after the fuel enters the furnace is very rapid. A difference that affects the control of combustion is the use of multiple pulverizers and the time delay for grinding the fuel. The variation in Btu content due to moisture and ash content variation creates additional control problems. The fact that the fuel as it enters the furnace cannot be measured directly creates another challenge for the control system designer.

21-1 Coal Btu Compensation

The various pulverizer control loops in Section 20 contain some form of coal flow measurement. In some cases this measurement is the speed of a volumetric or gravimetric feeder. In other cases a rough measurement of pulverized coal flow is obtained with the classifier differential pressure or pulverizer differential pressure. A total boiler fuel measurement is obtained by summing these pulverizer coal flow measurements. The accuracy in coal Btu of this summation is affected by the accuracy of the basic pulverizer signals, their linearity, and the variation in moisture and ash content of the individual coal streams.

An acceptable method of compensating for the potential inaccuracy of the signals and/or their summation is to use the boiler as a calorimeter. If the heat release of the boiler is known and the coal input is known, then the heat release can be used to continuously calibrate the coal flow measurement. This enables one to obtain a more precise fuel Btu input for use in the control system. A rough measurement of the heat transferred — heat release multiplied by boiler efficiency — is the boiler steam flow. The steam flow measurement alone may be no more precise than the rough coal flow measurement.

A precise measurement of the heat transferred must account for changes in steam pressure, steam temperature, feedwater temperature, and the heat taken from or added to boiler energy storage. For noncontrolled extraction turbogenerators, the energy input is proportional to the first-stage shell pressure of the turbine. If all the boiler output goes to such a turbine, the turbine first-stage shell pressure measurement is a suitable measurement of the boiler energy output. Under steady-state conditions, with the boiler pressure and temperature constant and the feedwater temperature constant, this value is proportional to the heat transferred in the boiler. The above assumes a nonreheat boiler and constant boiler efficiency.

If the feedwater temperature changes and thus alters the heat content in each pound of feedwater, the heat transferred is a different value. If the pressure and temperature are changing, the heat added to or taken from storage is changing.

A measurement logic arrangement that accounts for the total heat transferred and thus the total heat release is shown in Figure 21-1. The derivative or rate of change of drum pressure recognizes that the boiler drum and piping system has a fixed volume and the quantity of steam in that volume is approximately linear with respect to pressure. This is not correct over wide changes in pressure but over the relatively narrow range of operating pressure can be assumed to be true. A rate of change of this pressure is therefore equivalent to a rate of steam generation that should be added or subtracted to the rate of steam usage in order to account for the total steam generated.

If there is no single boiler-single turbine relationship, a different technique (as shown in Figure 21-2) can be used. In this measurement logic, the steam Btu flow is computed. The computation logic shown is identical to that of mass flow. The difference is that the functions of pressure and temperature shown combine the effects of Btu/lb and specific volume values.

The Btu flow is then multiplied by a function of feedwater temperature to obtain total energy transferred to the steam that is delivered to the steam system. This energy is then adjusted by the derivative of drum pressure to account for energy added to or taken from boiler energy storage. With constant boiler efficiency, the total is directly proportional to total heat release.

The total heat release values obtained by either of the above methods are, however, out of time phase with the coal flow measurement. The time delay between these measured values is the accumulated time for pulverizing the coal, transporting it to the burners, the combustion process, and the transfer of the heat generated. Any proper comparative use of these values must account for the time delay between coal flow measurement and the transfer of the heat.

The control logic by which the coal measurement is calibrated by the heat release value is shown in Figure 21-3. In this arrangement a summation (a) of the coal feeder speeds is used as the preliminary coal measurement. The calibration of this signal is effected in the multiplier (b). Essentially, the calibrated coal flow is compared to the total heat release in the delta (c). Any error between these two signals causes the integrator (d) to act until the signals are again in balance.

The time function (e) represents the delay discussed above and is adjusted so that the coal flow signal is in time phase with the heat transferred value. The signal from the time function is then an emulation of the actual heat release of the pulverized coal. It must be recognized that the integral (d) also has a time adjustment. This integral time must be slower than the time function (e) or the calibrating circuit shown will become unstable. The calibrated coal

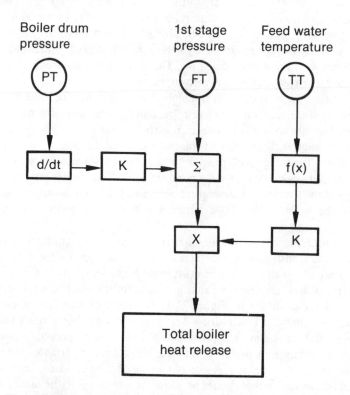

Figure 21-1 Computation Logic for Total Boiler Heat Release

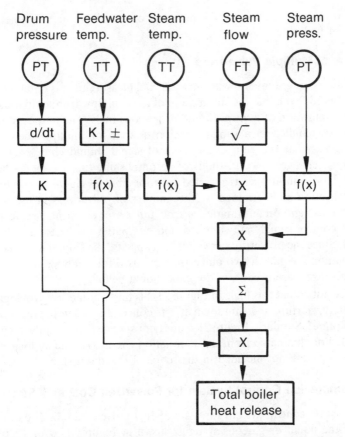

Figure 21-2 Computation Logic for Total Boiler Heat Release

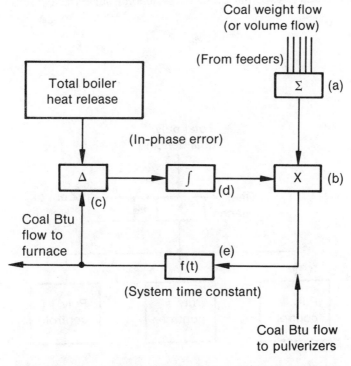

Figure 21-3 Calorimetric Calibration of Boiler Coal Flow

flow signal can now be used in the combustion control logic in the same manner as the measurement of fuel oil or gas.

21-2 The Use of Multiple Pulverizers

The preceding paragraphs show how a total coal measurement is obtained when multiple pulverizers are used. Note that the summation of the coal flow from individual pulverizers is similar to the totalization of multiple liquid or gaseous fuels. The output of the combustion control system is handled in a similar fashion as that for multiple fuels. The particular control problem is that the gain of the control signal should change as the number of pulverizers in use changes. Two methods of accomplishing this are shown below.

In Figure 21-4, a multiplier is inserted into the control signal to the pulverizers. The desired multiplication is determined by the number of pulverizers in use and this implies some sort of "in use and on automatic control" pulverizer counting circuit. In Figure 21-5, the pulverizer control signals are summed and compared to the total pulverizer demand control signal. Since the pulverizers are equal in capacity, the gain of their individual signals into the summation is equal. If two pulverizers are used for full boiler load, the gain should be 0.5; if 5 pulverizers are necessary, the gain should be 0.2.

It should be noted that this logic in Figure 21-4 is the "on-line" control logic and does not include the safety, startup, and shutdown effects concerning when the pulverizers should be counted as "on-line". Similarly Figure 21-5 does not show when the pulverizer control signal should be added or eliminated from the summation circuit. The safety logic and the startup and shutdown logic are not included in the scope of this discussion.

21-3 The Combustion Control System for Pulverized Coal as a Single Fuel

The combustion control system for a single-fuel pulverized coal-fired boiler can be an adaptation of the liquid or gaseous systems shown in Figures 17-9 or 17-10. In the same manner as the basic system is modified for various fuels or combination fuels, the tailoring points are those marked A, B, and C. Such a system for pulverized coal is shown in Figure 21-6.

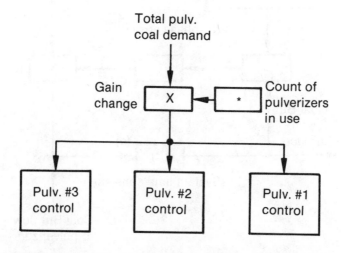

Figure 21-4 Gain Control for Number of Pulverizers in Use

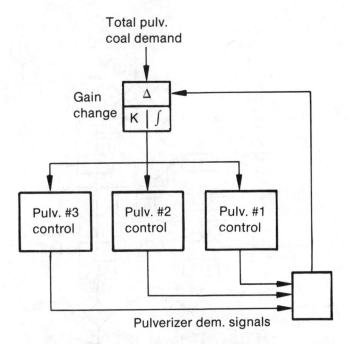

Figure 21-5 Gain Control for Number of Pulverizers in Use

The firing rate demand, the fuel controller, the fuel limit control, the air limit control, the air flow controller, and the flue gas analysis trim control are identical to those of Figure 17-9. The adaptation is that above point B the total coal measurement and its calibration circuit are connected. In addition, below point A the pulverizer controls are connected. The control station (a) is considered the pulverized coal master manual/auto station.

The basic pulverizer demand signal is developed as an output of the gain adjusting controller (b). The pulverizer loadings can be kept in balance with each other by the use of the bias-type manual/auto station that is part of the pulverizer block of control logic. The output of the individual manual/auto stations feeds back to summer (c) as part of the gain control circuit. The arrangement shown is for three pulverizers. If there were four or more, another pulverizer block would be used and the input gains of the summer (c) would be changed.

The time function (d) between the feeder speed summation (e) and the fuel controller (f) delays the proportional feedback to the controller and tends to provide an output with some derivative action. Derivative action in the control of a coal feeder causes the pulverizer coal level to change in a direction that will improve the pulverized coal response. If the feedback time funtion does not provide enough derivative action, the insertion of the proportional-plus-derivative function (g) or adding derivative action to controller (f) will supplement that already available.

The circled numbers in Figure 21-6 refer to blocks of control logic that are developed elsewhere in this text. The numbers are the section numbers where the detailed discussion is found.

An alternate approach to controlling a single-fuel pulverized coal-fired boiler is the adapted use of the steam flow-air flow system used for spreader stoker-fired boilers. Figure 21-7 is a variation of the parallel plus steam flow-air flow readjustment system shown in Figure 19-3. In Figure 21-7 the feeder speeds or other rough coal flow measurements are totalized in the summer (a) and the total fuel output is fed back to the fuel controller (b). A derivative function is added to this controller to temporarily increase the pulverizer coal level in order to improve the response of actual pulverized coal flow to the furnace.

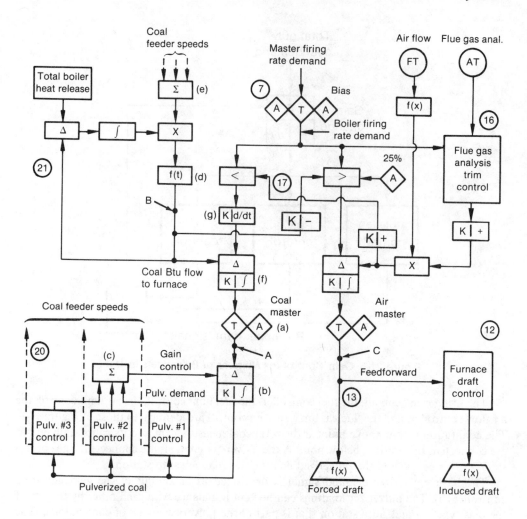

Figure 21-6 Control System for Single-Fuel Pulverized Coal-Fired Boiler

A preliminary position of the forced draft and associated furnace draft control is signalled by the feedforward of the boiler master firing rate demand signal (c) input to the summer (d). The other input to this summer is the final readjustment signal (e) from the total heat release-air flow controller (f). Steam flow can be compared against air flow as in the spreader stoker system, but performance is improved if the more precise total boiler heat release is used in ts place. The justification for using the more precise configuration is that pulverized coal-fired boilers are generally larger than spreader stoker-fired boilers. The incremental cost of the more complex control system in order to achieve more precise control of combustion can usually be justified.

Since the total boiler heat release signal includes the "put and take" of energy from storage as the boiler load changes, it is not limited to steady-state comparison to combustion air flow. In view of this, the time function (g) may not always be required. The compensation for variation in coal Btu value or other factors affecting the coal flow measurement is

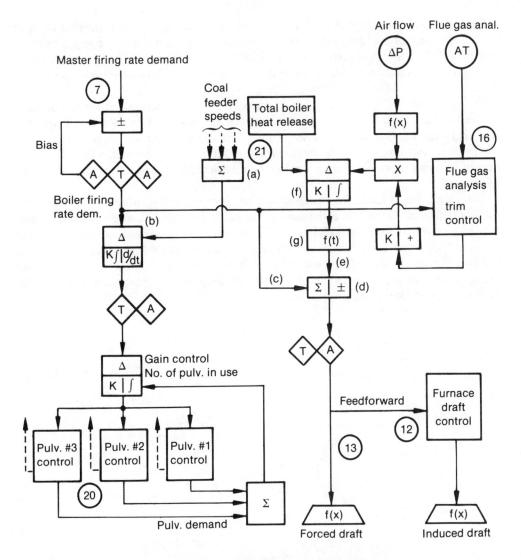

**Figure 21-7 Parallel plus Total Boiler Heat Release-Air Flow
Readjustment for Pulverized Coal-Fired Boiler**

accomplished by the integral function of controller (f). As in Figure 21-6 the circled numbers represent section numbers of this test where subsystem details are discussed.

21-4 Pulverized Coal in Combination with Liquid or Gaseous Fuels

When pulverized coal is burned in combination with liquid or gaseous fuels, the same burner air register assemby is used. The fuels have different nozzles but can be totalized on an air-required basis and fired together. The control application procedure is to add the coal flow to the flow of the other fuels to obtain a total fuel signal in terms of combustion air required. This is a modification above point B on Figure 21-6. In addition, below point A on Figure 21-6, the proper control actions to all fuels must be applied. Figures 21-8a and 21-8b show the modification to be made to the fuel control subsystem of Figure 21-6 when the boiler fires pulverized coal in combination with natural gas.

Since the total boiler heat release, as shown in Figure 21-8a, represents the heat from all the fuel, the heat from the gas portion of the fuel must be eliminated before comparison with coal flow. This is done by assigning a gas-fired boiler efficiency value in proportional function (a) and subtracting the result in the delta (b). The difference represents the heat release from the coal fuel.

The summation of rough coal flow in summer (c) can be compensated for Btu or other coal feed variations in the same manner described earlier. The total Btu values are then totalized in the summer (d) to provide a total fuel signal on an air required basis. The proportional functions (e) provide the small factors necessary for conversion from Btu value to combustion air required, difference in maximum Btu capacity, and difference in excess air requirement.

The output functions of the combination fuel system are shown in Figure 21-8b. The coal/gas ratio logic (a) and (b) is the same as the fuel ratio logic in Section 17. The proportional plus derivative function (c) is intended to accelerate the pulverized coal response of the pulverizers by temporarily increasing their coal level. It is inserted here rather than in the total fuel controller to prevent the derivative action from affecting the fuel gas or other combined fuel. The gain changes controls (d) and (e) compensates for the number of pulverizers in use and on automatic.

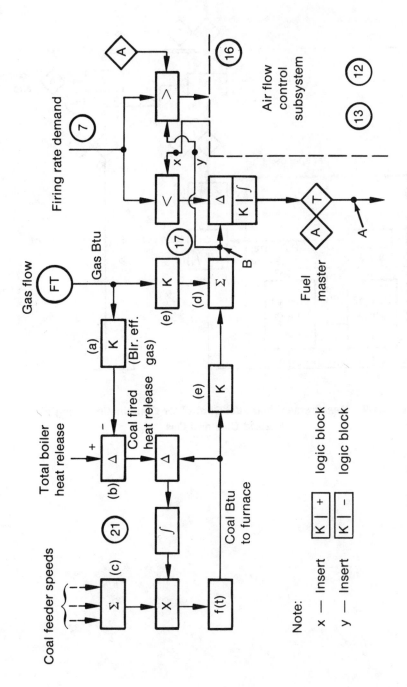

Figure 21-8a Measurement and Fuel Control for Combination Firing of Pulverized Coal and Gas

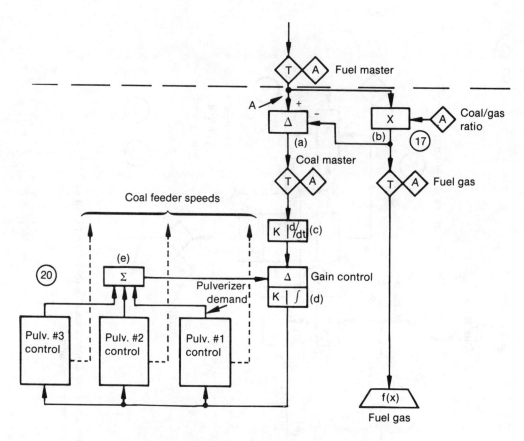

**Figure 21-8b Measurement and Fuel Control for Combination Firing of
Pulverized Coal and Gas**

Section 22
Combustion Control for Combination of Stoker and Liquid or Gaseous Fuel Firing

Some boilers are fired by using a combination of stoker firing of solid fuel and a separate burning system involving fluid or gaseous fuel burners. Typically, the solid fuel is a waste product such as wood bark, other wood waste, or solid refuse such as bagasse (the refuse from sugar cane), or other solid waste. The liquid or gaseous fuel may be burned as an auxiliary fuel to achieve a desired boiler steam flow, to temporarily replace the waste fuel, or to maintain ignition of the waste fuel.

The usual boiler air flow configuration is a single set of combustion air fans but with individual air flow control to the two sections of the boiler. In many installations the supply of waste fuel is intermittent, requiring the firing of auxiliary fuel. The intermittent supply of waste fuel may create periods when the stoker grate is bare and exposed to the radiant heat of the furnace.

All aspects of the boiler equipment arrangement and the operation needs must be considered when applying control equipment. Installations of this type are very individual and there is no one good control application solution to all the potential operation problems. There are essentially two separate control systems involved, and their manner of linkage determines the system results.

While the application solution may or may not be transferrable, the thought process required can be demonstrated by considering the needs of a particular installation. Assume a paper mill boiler that burns bark and wood waste with natural gas as an auxiliary fuel. The boiler is fired with a spreader stoker, with the gas burners mounted in the furnace wall above the stoker. The stoker and the gas burners have separate measurable and controllable combustion air systems. The combustion air is supplied by a single forced-draft fan with inlet vane control. There is also a single induced draft fan.

The boiler may be fired to full boiler rating on either fuel. The solid fuel is stored in a bin above the stoker and admitted to the stoker by variable speed screws on the bottom of the bin. Additional solid fuel is admitted to the bin with a constant speed belt conveyor that is not under the control of the boiler operator.

There may be times when the wood waste to the stoker diminishes or disappears through lack of supply to the storage bin or because screw feeders bridge over and do not supply fuel to the stoker. If there is no wood waste, the material on the grate will gradually be consumed and the bare grate will be exposed to full radiant heat of the furnace. If this happens, approximately 15% of full load air flow must be passed through the grates to avoid grate damage. The cooling air may be reduced by using water-cooled grates.

There are two desired operation modes. The first is that the boiler steam loading follows the demand of the steam system for steam in parallel with other boilers connected to the plant header. The second is that the boiler can be set at a given steam flow set point and maintain that steam flow even though the plant demand for steam may change. Figure 22-1 is a control system application that might be designed for such requirements.

The master (plant) firing rate demand is adjusted to the desired boiler firing rate demand in the bias-type manual-auto station (a). When on automatic, the output of this station will follow the changes in demand of the plant steam system by changing the set point of the steam flow controller (b). If the station is placed in the manual mode, the steam flow will be set by the operator changing the manual output of the station. The output of controller (b) is the gas-firing demand, which actuates a typical gas firing system as discussed in Section 17.

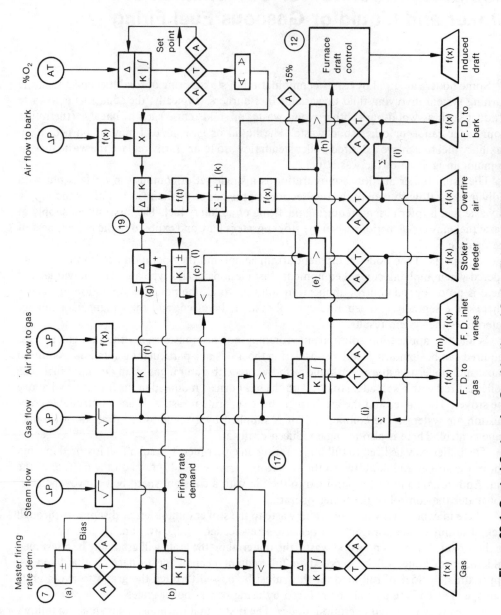

Figure 22-1 Control System for Combination of Stoker Firing and Firing Liquid or Gaseous Fuel

The boiler firing rate demand from station (a) is also the firing rate demand for the stoker. The high select (e) limits the minimum value of the signal to 15%. The low select (c) limits the maximum value to approximately 25% higher than the amount of steam generated from the waste fuel. The stoker firing demand signal is applied directly to the speed of the stoker feeder and is also applied as a feedforward to the summer, (k) which positions the combustion air flow to the bark or wood waste stoker. This part of the system can be recognized as the basic spreader stoker system shown in Figure 19-3.

The total measured steam flow can not be properly compared against only the air flow to bark combustion, so a means must be found to determine the steam flow that is generated by the wood waste material. This is accomplished in the functions (f) and (g). The proportional function (f) provides a scaling function that involves the capacity of the gas flow measurement, gas Btu/scfh, Btu added per pound of steam flow, and boiler efficiency in burning the gas. The output of the function (f) is the steam flow rate for the gas that is being burned. The gas-burning steam flow rate is then subtracted from the total steam flow rate to obtain the rate of steam flow generation from burning the wood waste.

To protect the grates from overheating, the high select (h) prevents the air flow through the stoker from being reduced below approximately 15% of maximum stoker air flow. The control signals to the forced draft dampers for controlling the combustion air to the stoker, and the gas burners are totalized in the summers (i) and (j). The resulting total is used to position the inlet vanes of the forced-draft fan (m) and to act as a feedforward signal to the furnace draft control. Another approach to the forced-draft fan control is to use a forced-draft duct pressure control loop.

The flue gas analysis control, % oxygen in this case, is set at a constant % oxygen set point. When the boiler is on gas firing and the stoker grate is bare, the air flow cannot be reduced because of the 15% flow through the stoker grate. For this reason the flue gas analysis control is applied only to stoker firing.

Assume that the grate is bare and the boiler is operating at 50% capacity on gas fuel. The signal from (g) is 0% since there is no steam generation from the wood waste. The firing rate demand signal to the stoker is at 25% because of the bias set in item (l). Stoker steam flow at 0 plus 25% equals 25% from item (l). This 25% is compared in the low select (c) to the total firing rate demand of 50%. At this time, bark flow is admitted to the storage bin. This feeds bark at a 25% rate; the grate is gradually covered, and the minimum 15% air flow causes combustion and steam generation by the stoker. The increase in steam flow causes the controller (b) to reduce its output. This increases the stoker demand signal by increasing the maximum limit in (c). This will continue up to the capability of the stoker or until the value of the boiler firing rate demand signal is reached. The fuel gas will reduce to its minimum firing rate or intermediate firing rate as called for by controller (b).

The stoker firing controls will predominate and follow the boiler firing rate demand as long as there is waste wood or bark fuel available. If there is insufficient stoker fuel, the gas fuel will increase to satisfy the boiler firing rate demand. Should the stoker fuel be reduced to 0 for any reason, the stoker will burn the remaining fuel on the grate, and the boiler will again be maintaining the boiler firing rate demand with gas alone as the fuel.

There are a number of potential solutions to this control problem. One solution uses closed-loop control on the stoker feed by totalizing the speed of the stoker feeders and using calorimetric continuous calibration of the resulting total fuel signal. Generally, more elaborate and precise control applications are used as boiler capacities increase.

Section 23
Atmospheric Fluidized-Bed Boilers

The term "fluidized bed" describes a process in which a bed of material is fluffed into a fluid mass by high velocity air or other gas that is applied to the under side of the bed. The effect is not unlike boiling water with the steam bubbles rising through the water and expanding the volume. For fluidized-bed combustion, the fluidizing air is the primary combustion air, with secondary air added as required to assure complete combustion.

The fluidized-bed process is not new and has been used for many years in the refining industry for fluid catalytic cracking and fluid bed hydrogenation and in other industries for such applications as coal drying and chemical vapor heating. A fluidized bed that operates essentially at atmospheric pressure is called an atmospheric fluidized bed.

Its use for boiler fuel burning is a somewhat newer process. Though a patent for the process was filed in 1944, fluidized-bed firing for boilers did not gain general recognition until the early 1970's. At that time the basic idea was for combustion to take place in the fluid bed and to obtain direct heat transfer by locating boiler tubes in the fluid bed.

A basic concept is that combustion temperature is reduced by mixing a large amount of noncombustible bed material with the fuel. The effect of the reduced combustion temperature is a significant reduction in NOx formation. If the noncombustible material is a sulfur-absorbing material such as limestone or dolomite, then sulfur in the fuel can be absorbed during combustion.

The material containing the absorbed sulfur is calcium sulphate that has become mixed with the chemically spent remains of the limestone or dolomite. This mixture of material is continuously withdrawn with the ash and replaced by fresh limestone or dolomite. Removal of the sulfur in this way eliminates the need for costly flue gas scrubbers that would otherwise be required for the elimination of SO_2 and SO_3 from the flue gases.

If sulfur capture is not required, the fluidizing material can be sand or a similar material. For such systems the material is not involved in the combustion process and does not require continuous removal from the bed. A small amount of makeup material to replace that mechanically removed with the ash or drained from the bed would be required.

A number of pilot installations were installed during the 1970's, with the fuel burning capacity of successive installations growing larger or using various combustible materials. One of these combustible materials was the long-term accumulation of coal mining waste called "culm", which contained a significant percentage of carbon. Fluidized-bed combustion was successful in recovering this wasted energy, and a number of installation have been made. Other installations were for industrial or municipal waste. During the 1980's as upscaling of capacity and new applications are being tested, the fluidized-bed boiler process is being applied commercially where such installations can be justified on an economic basis.

Atmospheric fluidized-bed boilers are generally of two basic designs. These are called "bubbling bed" and "circulating bed". Design variations may use elements of both of these. Circulating beds are lighter, with less material per unit volume, and are called "low density" beds. Low density is acquired by building up the fluidizing air pressure to achieve greater air velocity and expanded bed volume. Such a bed can also have a high density mode in which the fluidizing air is at a lower pressure and the material is fluidized to a lesser volume.

While a significant number of both types of fluidized-bed boilers have been installed, the process is still evolving. The control application practices are not firm and may be significantly influenced by the particular manufacturer's process design. The discussion in this section should, therefore, be viewed as a general guideline or starting point to be rationalized or modified in accordance with actual requirements.

23-1 Bubbling Bed Fluidized-Bed Boilers

A diagram of a bubbling bed fluidized-bed boiler is shown in Figure 23-1. The bed is located at the bottom of a stoker-type waterwall furnace with a significant portion of the steam generating tubes buried in the bed. Fuel is added by a screw conveyor, a gravity chute, pneumatic injection by an air stream of sufficient velocity, or a spreader stoker. Noncombustible bed material is added in the same manner as the fuel.

The bed is fluffed to the bubbling condition of a high density fluidized-bed by fluidizing air admitted below the bed. Since the bed is only bubbling, the flyash, or other carryover from the bed, is relatively low. In this respect the closest approximation is that of a spreader stoker. This carryover is cleaned from the flue gas at the end of the process and if significant amounts of carbon remain, it is returned to the bed.

Since the bed is in a bubbling and high density condition, changing the firing rate of an individual bed is limited to a maximum turndown of approximately 2:1. If greater turndown of the boiler firing rate is required, the boiler must use multiple bubbling beds. In this case, if a load of less than 50% is required, one or more of the beds would be slumped or allowed to settle by reducing the fluidizing air. Operation at the low loads may also require controls for alternately firing the individual beds.

As in the change of fuel supply to a spreader stoker, the burning carryover from the bed and the fine fuel particles burn in suspension and provide some immediate load-following response. Because of the large amount of unburned fuel inventory in the bed at all times, the major portion of load-following response results from changing combustion air flow. Changing the total combustion air relative to the fuel input affects the bed temperature and can be used to control it. Draining or adding material to the bed also affects bed temperature.

Several general statements concerning the basic operating control of such a boiler can be stated.

(1) The firing rate demand control and steam temperature control are no different from those of any other boiler and can be implemented in accordance with the discussion in Section 7. Typically, the firing rate demand change calls for the addition or reduction of fuel and fluidizing air in parallel.

(2) Fluidized-bed boilers are balanced draft and include both forced and induced draft fans. The furnace draft control and air flow measurement are the same as those of other boilers and can be implemented as discussed in Sections 11, 12, and 13.

(3) The feedwater control system requirements are identical to those of other boilers and can be implemented as discussed in Section 10.

(4) The key differences from other boilers lie in the fuel control and control of bed temperature. Bed temperature control may also involve the control of the ratio of fuel flow to combustion air flow.

If the fluidized-bed purpose is to capture sulfur from the fuel, it is particularly necessary to control bed temperature. The most efficient performance in sulfur capture occurs with the bed temperature between 1500°F and 1600°F. This temperature can be controlled by changing the mass of material in the bed or trimming the air flow to change the amount of fluidizing air with respect to fuel and other material being added to the bed.

At least one manufacturer suggests trimming the control of air flow from bed temperature as shown in Figure 23-2. If this is done, the result may be a bed temperature change but also a change in the fuel/air ratio. Such a fuel/air ratio change would be indicated by a change in the analysis of the flue gases. Another manufacturer suggests using bed temperature to trim the air flow control and, in addition, using the flue gas analysis to trim secondary air flow. The net result would be no change in total air flow but with a lowered amount of fluidizing or primary air.

An indication of bed temperature change also indicates an incorrect total amount of bed material or the amount of fuel in the bed relative to other material. If the problem is an incorrect total amount of bed material, then the solution is to drain material from or add material to the bed. If the bed temperature is high, adding more material to the bed and

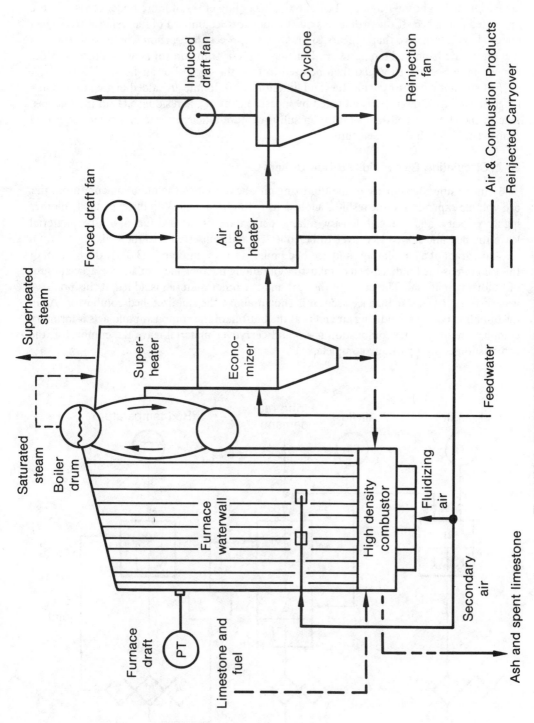

Figure 23-1 Bubbling Bed Fluidized-Bed Boiler

increasing its height may cool the bed due to the cooling effect of more boiler tubes that are immersed in the bed. If the problem is due to an incorrect amount of fuel relative to the bed material, an alternative that appears to involve less process interaction would be to trim the fuel input rate from bed temperature. Flue gas analysis trim control in accordance with the discussion in Section 16 would then be used to trim the control of air flow.

For sulphur capture systems, fuel and limestone or dolomite are added in a set ratio. On a longer-term basis, the ratio is adjusted by measurement of the residual SO_2 in the flue gas. Due to the effect of bed temperature on sulfur capture efficiency, such a system requires a proper control of bed temperature.

23-2 Circulating Bed Fluidized-Bed Boilers

By increasing the velocity of the fluidizing air above that of the bubbling condition, the bed volume expands and the result is a low density fluidized bed. With the expanded volume, higher velocity, and reduced density, a large percentage of the bed fuel and other material leave the bed and are carried over to be collected and reinserted into the bed.

A diagram of a circulating fluidized bed boiler is shown in Figure 23-3. At the bottom of this diagram is the fluidized-bed combustor containing the fuel and a relatively large amount of fluidizing material. The input of the fuel and the other material is added at this point. A waterwall heat transfer furnace section is shown above the fluidized-bed combustor. As in the bubbling bed, the fluidizing air enters at the bottom of the combustor and acts as primary combustion air. As the combustion gases and carryover material leave the combustor, the combustion is only partially complete.

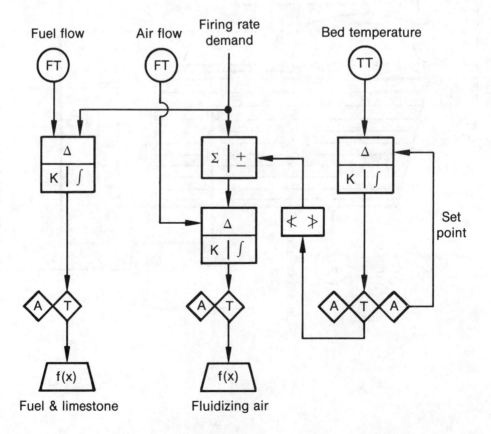

Figure 23-2 Bed Temperature Trim of Air Flow Control

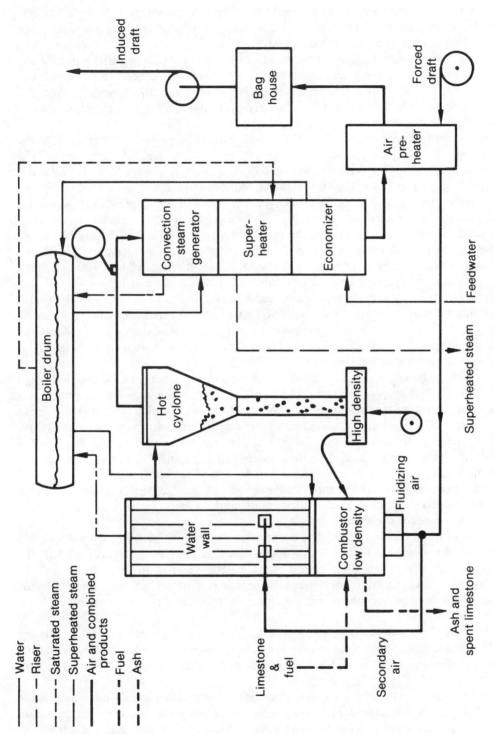

Figure 23-3 Circulating Bed Fluidized-Bed Boiler

The hot gases and burning particles leave the combustor, bounce off the waterwall heat transfer surface, and transfer heat by conduction, convection, and radiation before entering a hot cyclone or other type of particle separator.

As shown in Figure 23-3, the hot gases exit from the hot cyclone particle separator and then travel into and through the convection steam generating section. This flue gas then transfers additional heat in a superheater and/or economizer before passing through a combustion air preheater, induced draft fan, and particulate collecting equipment such as a precipitator or bag house.

In the arrangement shown, centrifugal force throws the heavier solid material to the outside of the hot cyclone collector and gravity causes these solid particles to fall. The material collected is then reinjected into the fluidized bed by gravity flow. To assist this gravity flow, a small bed at the bottom of the collector standpipe is lightly fluffed by high pressure air into a high density fluid bed. The static head of the material in the standpipe below the particle separator overcomes the difference in pressure due to draft loss and acts on this small high density bed, causing the material to flow into the low density bed of the combustor. Any fluidizing air of the small high density bed becomes secondary combustion air, with other secondary air admitted to the furnace above the fluidized bed. This circulation is continuous and can be increased or decreased by changing fluidizing air velocity and volume and thus changing the bed density and volume.

Controls for the circulating bed type of boiler are essentially the same as for the bubbling bed type. Some manufacturers may require controls for the circulation part of the installation, but others do not. As more material leaves the combustor, more is collected in the hot cyclone and returned to the bed.

As with the bubbling bed, the process is evolving and the manufacturer's control strategy should be the first approach to control application design. Steam temperature control and feedwater control are the same as for other boilers. Controls for firing rate demand differ to some extent. The circulating fluidized-bed boiler can be turned down over a greater range, and to obtain greater turndown than the 2:1 available with bubbling beds, a fewer number of beds are required. This greater turndown availability eliminates the requirement for any controls to alternately fire the multiple beds under a low load condition.

The furnace draft control balance point is shown on Figure 23-3 and the suggested initial set point value is –0.5 inch of H_2O. Furnace draft control application as discussed in Section 12 is appropriate for the circulating fluidized-bed boiler.

Most of the factors concerning bed temperature control with bubbling bed boilers apply equally to circulating bed boilers. An exception is that the circulating bed does not have the capability of bubbling beds to be cooled by adding bed material. In addition, the fire side waterwall outlet temperature, which includes the solid material temperature, can be affected by the distribution between the primary fluidizing air flow and the secondary air flow that is added above the bed.

Because of the much larger amount of material returned to the bed with the circulating bed boiler, the temperature of the solid material that is returned also affects bed temperature. One manufacturer uses this fact in the process design and the bed temperature control strategy.

With the number of factors that can affect the bed temperature and the differences in the way these factors interact with different boiler designs, some period of testing and evaluation of different process and control strategies will be necessary to solidify design practice.

Section 24
Control System Complexity and
Future Directions for Boiler Control

This text deals with those aspects that most relate to the control of boilers and to the control application that results. The control application is essentially independent of the particular set of hardware and software with which the control application is implemented, whether pneumatic, electric analog, or digital. At this time the hardware/software revolution is virtually complete, and over 90 percent of future boiler control systems will likely be implemented with digital control.

The use of digital control will facilitate the integration of the on-line control (covered in this text) with the digital logic for startup, shutdown, and safety monitoring that is not covered here. Another particular benefit of digital control is the non-drift characteristic of the control system tuning and alignment.

The use of digital control also makes much easier the implementation by software of more complex control algorithms that require the expense of considerable additional hardware when implemented with analog control.

This advantage is of considerable benefit in the size range of industrial and small electric utility boilers covered in this text, but is of particular benefit to the control of the larger and more complex electric utility boilers. While the scope of this text does not allow a detailed discussion of such utility boiler control, several key areas of such control are identified in the following.

24-1 Complex Areas of Electric Utility Boiler Control

In almost all cases, the configuration of a modern electric power generation unit consists of the combination of a single boiler feeding steam to a single turbogenerator. Complexity of the control system is one result of the use of multiple pulverizers, multiple boiler feed pumps, and multiple forced and induced draft fans for the single boiler unit.

The total boiler control system must also include operational restraints resulting from the failure or shutdown of part of the multiple boiler auxiliaries above. Such restraints are called runbacks. For example, if the unit needs two boiler feed pumps in order to safely maintain full load, control circuits must be in place to run back the boiler-turbogenerator load to a single pump load capability should one of the two pumps fail or be shut down. The load runbacks are applied to the firing rate demand control section of the total boiler control system.

A third area of control complexity is the result of some of the safety considerations for these large units. An example is the implosion control section of the system. Implosion is the result of high negative pressures acting on large area expanses of the boiler walls or flue gas ductwork. The addition of flue gas scrubbers to utility boilers results in the need for induced draft fans with a high negative pressure potential. The sudden reduction in furnace temperature that arises from a main fuel trip causes an immediate loss of flue gas flow and high negative pressure throughout the boiler unit. In such an emergency, implosion controls are necessary. Implosion control is an overriding set of control logic that runs back the position of the induced draft fan dampers and takes other appropriate action relative to the forced draft.

Since there is a single boiler-single turbine system, the steam pressure used is normally the pressure at the turbine throttle. In these units, a change in firing rate demand begins with a change in electrical load demand. This demand is signalled to the power plant by a central power dispatcher who requests a load increase from the plant. If the load is changed manually, an operator uses a raise-lower switch to operate a motor- operated speed changer in the turbine governor control circuit. The turbine valves change position and the steam flow to the turbine changes. The result is a change in the turbine throttle pressure.

It is necessary that the system include rate-of-change limits to avoid unsafe, unstable, or potentially damaging actions to the boiler and/or turbine. A knowledgeable operator keeps this in mind as he changes speed manually. A large percentage of present-day units include automatic dispatch systems, and this requires the inclusion of automatic rate of change limits. Electric system frequency affects generator load and is also included as an input to the unit load demand control. A simplified diagram of the unit load control development is shown in Figure 24-1.

There are three basic methods of generating the firing rate demand control signal. A digital (usually distributed microprocessor-based) system often incorporates all three of these as operation modes for operator selection. The three methods are described below.

(1) Boiler following

In this arrangement the turbine load is changed by the operation of the turbine steam valves and the resulting steam pressure change causes the boiler firing rate demand to change. For a load increase, the drop in steam pressure results in borrowing from boiler energy storage to assist in furnishing the desired load. The pressure will drop to the level necessary to satisfy the steam flow demand. This arrangement is the most responsive but the least stable of the methods of firing rate demand development. A simplified diagram of a boiler following arrangement is shown in Figure 24-2.

(2) Turbine following

In this arrangement steam pressure is controlled by the turbine valves. A change in the electric power requirement causes the boiler fuel and combustion air to change. The resulting change in turbine throttle pressure causes the turbine steam valves to modulate and control the turbine throttle pressure.

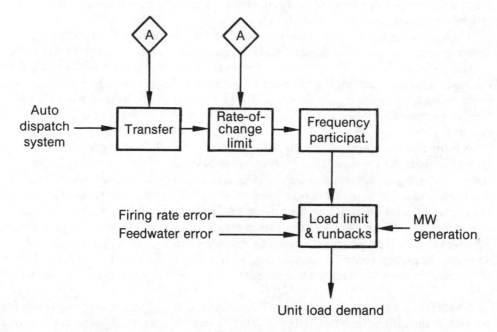

Figure 24-1 Unit Load Demand Control Development

There is no electric load response until the turbine throttle pressure has changed as a result of the change in firing rate. This type of control is very stable due to the stablility of the turbine valve control of the turbine throttle pressure. Very little energy from boiler energy storage is used due to the close control of turbine throttle pressure. With no borrowing or repaying of energy from boiler energy storage, the response is slow and due solely to the time constant of the boiler process. A simplified diagram of the turbine-following arrangement is shown in Figure 24-3.

(3) Coordinated Control

The use of the term "coordinated control" here is intended as a generic term for combining a large part of the responsiveness of a boiler-following arrangement with a large part of the stability of the turbine-following strategy. In such a control strategy, inputs of electric load, steam flow, turbine valve position, steam pressure, and feed water temperature are brought together into a control logic subsystem that controls both the turbine throttle steam control valves and the firing rate of the boiler. The system is tuned to allow a limited amount of turbine steam valve change balanced against a limited amount of turbine throttle pressure change. The limits set are those best meeting the elctrical load demand, unit safety, and overall boiler, turbine, and control system stability. A simplified diagram of such a coordinated control is shown in Figure 24-4.

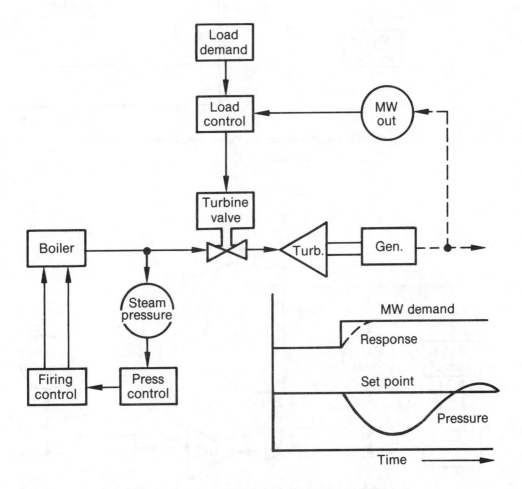

Figure 24-2 Boiler Following Firing Rate Demand Control

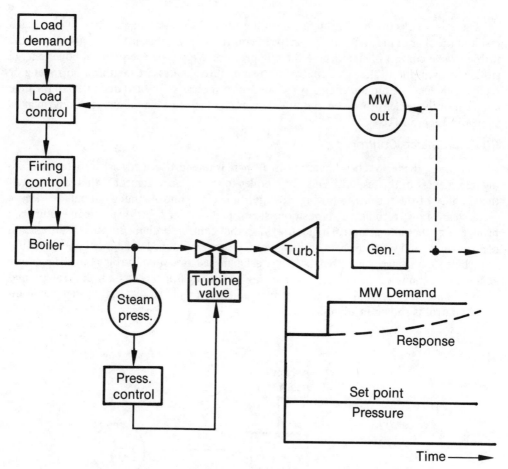

Figure 24-3 Turbine Following Firing Rate Demand Control

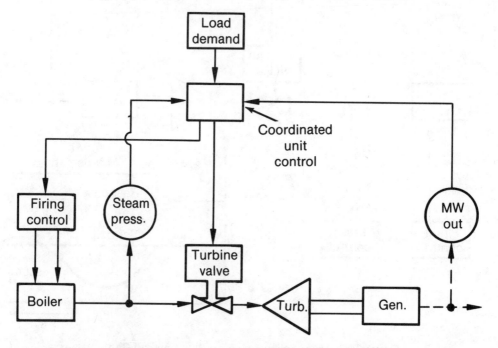

Figure 24-4 Coordinated Control of Boiler Firing Rate and Turbine Steam Flow

A fourth arrangement, which is a recently introduced variation of boiler following, should be mentioned. This strategy, which is shown in the simplified diagram of Figure 24-5, computes the firing rate demand signal rather than using the output of typical feedback or feedforward-plus-feedback control of turbine throttle pressure. While the author has had no experience in using this type of control, it appears to offer all the responsiveness advantage of the more common arrangements but with greater overall system stability.

As shown in Figure 24-5, the turbine first stage pressure, a measure of the turbine throttle input energy, is divided by the actual turbine throttle pressure to develop a linear pseudo turbine valve position signal. This signal is then multiplied by the turbine throttle pressure set point to develop the basic firing rate demand signal. If the actual pressure is lower than the set point, the calculation produces a firing rate demand that is higher than the calculated pseudo turbine valve position. The opposite is true if the actual pressure is higher than the set point.

The steam temperature control of large utility boilers is considerably more complex, may include the control of dual furnaces, and is more interactive with the steam pressure control than the basic steam temperature control discussed earlier in this text. The system usually

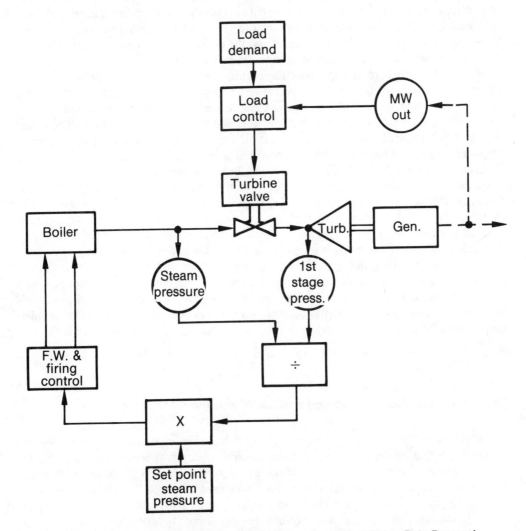

Figure 24-5 Variation of Boiler Following — Computed Firing Rate Demand

includes the control of the reheat temperature of steam, which issues from the turbine after partial expansion, is brought back to the boiler, reheated, and again admitted to the turbine. Main steam temperature and reheat temperature must be coordinated, and each may utilize two or more control devices such as spray water, flue gas recirculation, or burner tilts.

The furnace draft control must incorporate implosion protection and multiple forced and induced draft fans are usually involved. Many large utility boilers include flue gas scrubbers; their action can interact significantly with the control of furnace draft or the control of air flow.

The control application design of such units is today based on the skill and intuition of experienced utility boiler control application engineers. The use of what is known as "modern control theory" methods to solve the complex and interactive control problems of these large multivariable systems has not been used to any significant extent. The successful use of such methods relies on the development of and inclusion in the control system of an accurate boiler model that runs significantly faster than its real-time counterpart.

In one form of such systems, the boiler model is connected into the control system as an observer with the same set of process inputs as the basic system. A change in the inputs results in a rapid computation of model outputs that predict what the process measurements of the real system will be, based on the control actions being applied. Any deviations between the control system set point values and the predicted future process measurement values are then used to correct the control actions before the predicted deviations appear as real deviations. If the process model is accurate and the overall system is properly designed and tuned, more stable and precise control performance is obtained.

With such a system, modeling errors may require the basic system to carry the additional load of undoing incorrect control actions that result from incorrect predictions of the model. Similarly, errors in process measurements to both the model and the basic system may significantly reduce the benefit that could be obtained from an accurate model. The model should, therefore, be able to learn and make itself and any incorrect measurements better. A simplified diagram of such a model incorporation is shown in Figure 24-6.

Such utility boiler control system designs are being tried in a relatively small number of cases, and encouraging results of improved precision and control performance are being obtained. The expense of the model development and the lack of system understanding by the typical control systems engineer, who must take care of the regular day-to-day operation of the system, are two key drawbacks to the use of this technique.

Such methods will not be used unless it can be shown that present control performance standards result in losses that would pay for the cost of improved control performance. It is not enough that steam pressure, steam temperature, or fuel/air ratio be performed with greater precision. It is important to note that such systems would have been practically impossible without the use of digital control. More use of such systems can be expected as they become better understood and their implementation can be made at lower cost.

24-2 Improving Control Precision and Stability without Modeling

While the use of modern control theory and its inclusion of accurate boiler models is a developing technique for more precise boiler control, there are other avenues available for improving control performance. Some of these have received little or no attention by boiler control application engineers while others have been used sparingly. These arrangements do not require costly boiler system modeling and are more easily understood by the typical user's control engineers. Digital control offers the opportunity to take full advantage of these techniques, though some of them can easily be implemented with analog control.

24-2-1 Self-Tuning or Adaptive Tuning

Self-tuning controller algorithms are now available for insertion into control systems. Such controllers automatically compensate the controller tuning as process or boiler conditions change. Adaptive tuning can also be implemented from load or some other variable of the process. The technique is shown in the diagram of Figure 24-7. Its use is demonstrated in a control system in Figure 10-17 of this text.

Another candidate for the use of this type of control improvement is in boiler steam temperature control. It is known that the optimum tuning of the gain and integral modes of the steam temperature controller changes as load on the boiler changes. Some software controllers in microprocessor-based systems allow the direct input of an external signal for adaptive controller tuning. If not, the objective can be accomplished by the combination of standard control algorithms that is shown in Figure 24-7.

24-2-2 The Calibrating Integral Circuit

An example of this is shown in Figure 7-21 of this text. This technique configures the proportional and integral functions in separate controllers. A long-term proportional error causes the integral controller to act in a manner that will correct misalignment in the gain of a feedforward or measurement signal. Another example in this text is shown in the fuel Btu compensation circuit shown in Figure 21-3.

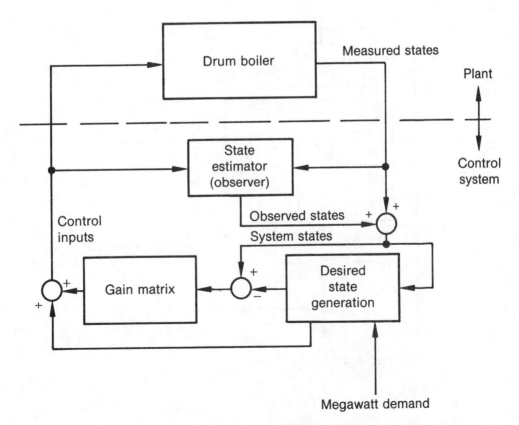

**Figure 24-6 Functional Diagram of Fossil Plant Modern Control System
Including an Observer Model**

24-2-3 Computation of Unmeasurable Variables

A typical example of the use of this technique is the computation of boiler heat release. Boiler heat release cannot be measured directly, but it can be computed by using the boiler as a calorimeter. One form of such a computation is shown in Figures 21-1 and 21-2 and is discussed in Section 21. An addition to this computation that would further enhance its precision in the control circuit is another computation representing variations in boiler efficiency.

The text also describes how the computation result can be combined with a calibrating integral circuit to improve a relatively poor measurement of boiler fuel input. This is shown in Figure 21-3 of this text. For control system performance improvement, the control engineer should investigate whether available measurements can be combined in a computing circuit and the calculated result used in place of an unavailable direct measurement.

24-2-4 Cross Coupling

The performance of many control systems is compromised by the interaction between the various sections of the overall system. In boiler control the interaction is typically among the feedwater control system, the combustion control system, and the steam temperature control system.

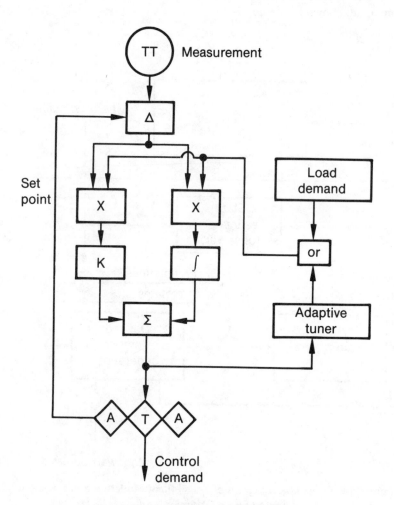

Figure 24-7 Adaptive Controller Tuning

Assume a utility boiler operating at 2400 psig and 1000° F. Assume also that steam temperature is controlled with a water spray and firing rate is controlled from steam pressure. At a given point in time the steam pressure may be 2425 psig and the steam temperature 985° F. The conventional action would be to reduce firing rate in order to reduce steam pressure to the 2400 psig set point and to reduce spray water flow in order to increase the steam temperature to the set point of 1000° F.

Since a reduction in firing rate would further reduce steam temperature, the result would be a further decrease in spray water flow. Spray water flow is added to the boiler steam generation and the decrease in spray water flow would tend to further decrease steam pressure, thus causing an increased firing rate and subsequent pressure increase. The result is that the two control subsystems would interact and fight each other. Precision of control would be reduced.

An analysis of the thermodynamic steam and water properties shows that the heat content of the steam is due almost entirely to the steam temperature. Steam temperature is also the major factor in determining the specific volume of the steam. At 2400 psig the pressure values used in the example would have very little specific volume or heating value effect. For a given steam flow, the specific volume of the superheated steam, along with the volume flow, determines the steam pressure in the confined volume. In essence, pressure can change because of load changes, temperature changes, and spray water changes. Most systems consider only load changes.

Both pressure and temperature should, therefore, be used in such systems as inputs for controlling firing rate, and both pressure and temperature should be used for controlling the spray water flow. By cross-coupling the steam temperature with appropriate gain into the firing rate control and steam pressure with appropriate gain into the spray water control, the overall system performance is improved.

Using the previous example , the 2425-psig pressure controller would reduce firing rate a smaller amount than before due to the increased firing rate request from steam temperature, with the remainder of the pressure reduction needed coming from reduced spray water flow. The steam temperature controller would call for a reduction in spray water flow. While in the right direction to also reduce the steam pressure, the amount of the change in spray water flow would be tempered by the steam pressure input.

Assume that the pressure is 2425 psig and the temperature is at the set point of 1000° F. Both the firing rate and spray water flow would be decreased by the steam pressure controller with no action from the steam temperature controller. If the system is properly tuned, the firing rate decrease would have the effect of reducing steam temperature and the spray water flow decrease would have the effect of increased temperature. The result is little, if any, effect on the steam temperature.

24-2-5 Nonlinear Control

There are several varieties of nonlinear control. One common method is the "error squared" algorithm in which the proportional error from set point is squared before the controller gain is applied. A nonstandard nonlinear gain can be applied to a measured linear deviation from set point and can be inserted into a control loop in the manner shown in Figure 24-8. In this case, the desired gain is computed as a function of controller error and can include a different error-gain function, depending on whether the set point error is positive or negative. The desired gain can be a function of other relationships as well.

A particular application of nonlinear controller gain is to make possible the use of the plant steam system as a steam accumulator. In some plant systems, sudden large but temporary energy demands may overtax the boiler system in meeting these large load swings. If a nonlinear steam pressure controller, as shown in Figure 24-8, with controller gain as a function of steam pressure is used, then the steam system energy storage can be used to assist the boiler in meeting these sudden large steam demands while changing the boiler firing rate characteristic to a more desirable pattern. Such a system would allow the boiler steam pressure to decay or build up at programmed rates. If this were done, it would be necessary to pressure-compensate the measurements of boiler steam flow and boiler drum level.

24-2-6 Artificial Intelligence or Expert Systems

Self-tuning controllers are one aspect of this technique that inserts the methods of the expert individual into the control system. The expert control engineer observes the action and performance of the system before he makes tuning and alignment adjustments or reconfigures the system. In making his decisions he uses certain thought processes that only he can identify. If these thought processes and techniques are properly inserted into the system as software, the system will be able to diagnose its ills and administer the proper remedies. Manufacturers of control systems are presently experimenting with this technique.

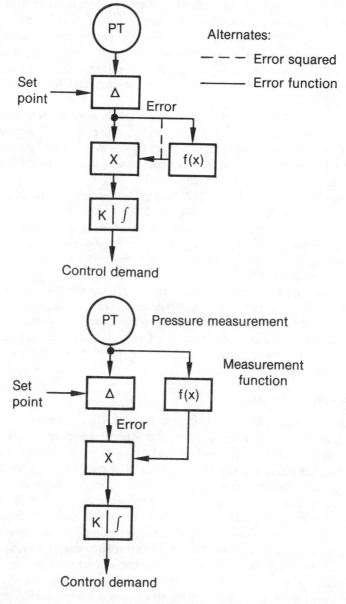

Figure 24-8 Nonlinear Control with Controller Gain a Function of a System Variable

24-3 General Observations Relative to Boiler Control Application

Boiler control application should be undertaken only by those individuals who have a thorough knowledge of the combustion and steam generation process and a thorough understanding of process control. The process of solving a boiler control application problem is similar to that of solving any other problem. There are three steps that some authorities have given weights of 50 percent, 40 percent, and 10 percent.

(1) Fully understanding and stating the problem — 50 percent.

To properly state the control problem, all of the objectives of the control system, including the necessary control performance, must be clear. An objective of plus or minus 5° F for steam temperature control should take into account the economic benefit of plus or minus 5° F as compared to plus or minus 10° F or plus or minus 15° F. If control systems and how different control elements interact, or the manner in which the different elements of the process act and interact, is not fully understood, the control problem cannot be stated correctly.

(2) Determining various alternative solutions to the problem — 40 percent.

After the problem has been properly stated, there may be several potential solutions with each having advantages and disadvantages. Proper potential solutions cannot be developed without the process and control system knowledge.

(3) Selecting the best alternative solution and implementing that solution — 10 percent.

This step also requires a thorough knowledge of both process and control systems.

The control solutions should be based on the functions needed rather than the control means. The economic justification may require a simpler, lower cost solution, but this should not get in the way of the proper analysis of the problem. Though digital control is now the predominant control means, analog control may be the proper solution to a problem involving the matching of existing equipment, spare parts, and the retraining of operation or maintenance personnel.

With the advent of digital control, changes in other control application practices may also begin to emerge. In the past the boiler control functional separation has been based on a breakdown that was based on the modulating functions of combustion control, steam temperature control, feedwater control, the digital functions of burner management, the digital functions of burner light-off and flame safety control, and the digital interlock functions. The result has been that each installation is a custom application.

Since microprocessor-based systems can handle modulating and digital logic functions equally as well, it appears that simpler coordination and greater standardization of application can be made if we depart from the functional control system breakdown and organize the system around unit controllers for the type of equipment being controlled.

An example of such a controller would be a pulverizer controller. All of the modulating functions, the digital interlocking functions, the startup and shutdown of the burners, and all of the safety function would be a part of that pulverizer controller. Other pulverizer controllers of that installation would be identical. A single demand for more or less pulverized coal would be sent from a master system and the pulverizer controller would perform all necessary functions to supply the coal. This would include starting up or shutting down the pulverizer and all its auxiliary devices, if necessary. Other possible unit controllers would be the induced draft controller, the forced draft controller, feedwater pump controllers, etc.

Boiler control has made tremendous strides in the past 50 years as we have progressed from direct-connected firing aisle control panels and direct-connected regulators to the distributed microprocessor-based systems of today. With all the power and flexibility of modern distributed digital control at our disposal, the boiler control application engineer now has all the necessary tools and the opportunity to use them.

In the future, the many new and more sophisticated boilers, turbines, and energy systems will require the design and implementation of control systems. Add to that the prospect of replacing 80 to 90 percent of all the boiler control systems that have been installed in the last 30 to 40 years.

What a wonderful time to be a boiler control application engineer and have a part in the next generations of progress.

Index

About the Author

Sam Dukelow's career has been dedicated to improving efficiency of energy conversion through designing, developing, and using energy-related systems.

A licensed professional engineer and President of Energy Conservation Services in Hutchinson, KA, Dukelow has conducted more than 500 combustion tests on boilers and other combustion processes in utility and industrial plants, and has tuned more than 300 automatic combustion control systems.

His expertise in fuels and controlling combustion ranges from solutions for new unit applications to retrofitting existing installations to save fuel, to change fuels, or to improve overall control.

This expertise is backed by 38 years of experience with Bailey Controls Company, a subsidiary of Babcock and Wilcox Company.

Dukelow has applied his experience in testing combustion and energy utilization to simplifying boiler performance monitoring techniques and methods. His approaches are easy to understand and use effectively.

He has lectured at energy conservation seminars in many colleges and universities. He coordinated and acted as technical adviser for the award-winning five-film series on Boiler Control, produced by the Instrument Society of America.

Dukelow has written numerous magazine articles, technical papers, and handbook sections. Among those are (Chilton) "Instrument Engineers Handbook," (McGraw Hill) "Handbook of Energy Technology," the National Bureau of Standards "Handbook # 115," (Kansas State University and ISA) "Improving Boiler Efficiency," as well as two software packages for ISA: "Boiler Efficiency Calculations," and "Thermodynamic Properties of Water and Steam."

He earned a bachelor of science degree in mechanical engineering at Kansas State University. He is a member of the American Society of Mechanical Engineers and a Senior Member of the Instrument Society of America.

His awards and recognition include the 1979 ISA Power Division Annual Achievement Award for outstanding contribution to the advancement of instrumentation in the power industry.